Volcanoes, El Niños, and the Bellybutton of the Universe

Volcanoes, El Niños, and the Bellybutton of the Universe

D.A. Walker, Ph.D.

This book was printed in the United States of America.

Cover photos by the author

Illustration by Brooks Bays

To order additional copies of this book, contact:
Xlibris Corporation
1-888-7-XLIBRIS
www.Xlibris.com
Orders@Xlibris.com

Contents

Long separated in space and time from their Mother, the basaltic rocks shaped by an unknown spirit seemed all-knowing. There in the quarry crater of Rano Raraku, with the sunrise, a full-moon, and an ancient comet overhead, I wondered what those silent figures would know if they had minds and memories to match their mysterious faces. I wondered whether any secrets of their origin or their Mother would ever be revealed to me.

Acknowledgments

Thanks to all the leaders, teachers, colleagues, secretaries, editors, reviewers, friends, and relatives who were a help and inspiration for this work. Thanks also to the funding agencies, their program managers, and individuals whose tax dollars support scientific research.

Preface

Several years ago, a publisher asked if I would be interested in writing a book. He had read about the controversy surrounding my idea for a possible cause of El Niños in the *New York Times*, and thought the controversy and my experiences could be an interesting story. I said *"maybe,"* but was skeptical. I had never written a book, nor did I know why I should want to. Besides, why would anyone read such a story. Answers to those questions weren't obvious, so the idea was put on hold while I wrestled with other research topics. In the years that followed, newspaper and magazine articles on my work, as well as a brief appearance on CBS's *48 Hours*, convinced me to write the book. There was so much more to the story than could be expressed in telephone interviews for the newspapers and magazines, or in the 30-second sound bite on *48 Hours*. Adding to my motivation was the formation of a powerful new El Niño and equally powerful evidence of its birth.

The adventure of science is a truly wonderful experience, and the stories of those adventures should be shared so as to encourage and inspire future generations. Scientists are mankind's greatest detectives. Good science requires hard work, imagination, an open mind, persistence, a spirit of adventure, and an "attitude" (i.e., you *can* do anything you want and you *can* solve any problem). With dedication to those characteristics, pieces of "our puzzles" will continue to reveal fascinating and unimagined pictures of the natural world.

Moon Over Miami

It was March 1970 and here I was on Easter Island, the extreme limit of Polynesian expansion, just a few short weeks after my wedding to Francine, my Hawaiian lover. I was met by Bernardo Blas who had been on the island before me on several occasions to construct the seismic vault for what would be one of the most remote seismic stations on the face of the earth. Easter Island is in the middle of the southeastern Pacific—more than 2000 miles west of Santiago, Chile and more than 2500 miles east of Papeete, Tahiti.

I had some experience with the trials and tribulations of remote installations having previously established similar stations at Midway, Wake, and Marcus islands in the Western Pacific as a graduate student at the University of Hawaii. Bernardo, who worked out of the University of Chile, was anxious to see the components that I had brought from Honolulu. We both hoped that the data acquired by our new seismic station would provide a better understanding of the earth's structure under deep ocean basins. The crates of instruments were off-loaded from the Boeing 707 that had only recently begun service to Easter by way of Santiago and Papeete. There was no terminal, just a rectangular alignment of empty 55-gallon drums on the tarmac to define the passenger "lobby". By the time we had the crates off the runway and into the seismic vault, it was dark. As we walked back towards the house that Bernardo had rented, I asked him: *"Is there a restaurant somewhere?"* He didn't say "yes" or "no". He just laughed. I didn't press the issue. Walking in the full-moon and passing through the shadows of the famous giant statues, I had a flashback to the earlier full-moon that had brought me to this place.

I saw that earlier moon when I was six or seven, riding down the Vermilion River in the back cockpit of my grandfather's Gar Wood speedboat. Grandpa owned a boat yard next to a well-known riverside restaurant called McGarvey's. Vermilion is a small resort town on Lake Erie between Cleveland and Sandusky. In those days I would spend my summers with my grandparents in that magical place—far from the south side of Chicago where I lived in an apartment with my parents and older brother. In Vermilion I was a happy "Tom Sawyer" terror. I had a bike and my own "yacht"—an eight-foot dinghy with a 2½ horsepower Evinrude engine. Vermilion even had an amusement park called Crystal Beach. It had a roller coaster that was always falling apart and a lot of other good rides that would make you lose your lunch or dinner, depending on the time of day. Occasionally, we would even go to a much larger amusement park called Cedar Point near Sandusky. I roamed everywhere in Vermilion and had many close calls with cruisers and speedboats. There were frequent swampings of my boat, appropriately named after my favorite character—"Sinbad the Sailor". By some miracle I managed to survive those summers, not realizing at the time how dangerous my adventures were.

Getting back to the full-moon on the river, I had, as usual, rushed out of the bushes and jumped into the back of the speedboat as it pulled away from the dock. My grandpa always pretended that he didn't see me, but knew my routine. After all, it was part of the show. Most of the time the paying customers didn't notice because they were facing forward in the two cockpits ahead of the engine compartment. Also, I was light, would land on the rear cockpit seat, and immediately drop to the floor. I would stay hidden until we had traveled too far for my grandpa to turn back and kick me off, even if he really wanted to. When

I finally popped up, the customers would be shocked to see me in the back of the boat. My grandpa would say something like: "That darn kid. I told him not to do that. He's gonna get it when we get back." After a high speed ride on the lake with much concern for the little boy in the back—he was being drenched by spray flying over the back of the boat and might even be thrown out in some of the tighter turns—we would return to the harbor for a leisurely cruise down the river and its lagoons. At that point I would ask the customers if they would like me to sing them some songs—a dime apiece or three for a quarter. Although I only had three songs, I would sing them as many times as they liked, so long as they paid. The songs were *Cruising Down the River*, *I'm Looking Over a Four Leaf Clover*, and *Moon over Miami*. On that particular night, after singing *Moon over Miami*, I noticed that the moon over Vermilion was also full. I thought of how summer was coming to an end and how I dreaded returning to Chicago. I thought of how my grandpa and grandma closed the boatyard in winter—the lake and river were frozen—and spent their winters in Florida. I thought of the fun and beauty of summers and how smart my grandparents were to go to a place where it was summer in winter. I decided then, under that full-moon, that I would live where it was summer in winter. As luck and/or determination would have it, I made it to Hawaii. Now on Easter Island, I was in another tropical place where its full-moon was reminding me of that earlier moon and a dream that came true.

When we arrived at the house, I realized why Bernardo had laughed when I asked him about a restaurant. There weren't any! But there in the kitchen were some huge lobsters that he and some locals had caught the night before. Dinner was no problem, or as the locals would say: "Ena problema".

The next day Bernardo watched in amazement as I unpacked the gear and had the station up and running in a few hours. Things went better than we expected. Since there was only one plane a week, what then were we to do with all our free time? "*Ena problema!*" We could spend a few minutes every day checking the station and then spend the rest of the time exploring the island. Or, we could track down the local artists and barter for their famous wooden carvings and stone sculptures. The next day Bernardo had a few more things he wanted to check out at the station, so I went off to the village in search of souvenirs. I had brought a bag full of fishing supplies and tools, having been told that such items were of greater value than money. It was fun, hiding the contents of the bag, and slowly, incrementally adding to the "kitty" for a particular carving of interest. The artists couldn't speak English very well and I couldn't speak Spanish at all. Nonetheless, we had little trouble communicating. The negotiations were friendly, interesting, and thoroughly enjoyable. At the conclusion of the transactions, everyone seemed pleased with the exchanges.

In the days that followed, we explored everywhere. The island and its statues were hauntingly beautiful. I thought that if there were such things as ancient astronauts, they probably had landed here. At night the "chicken skin" (i.e., the Hawaiian equivalent of "goosebumps") feeling of spirits and mystery were especially powerful.

Just after the station became operational and while the moon was still full, we spent an entire evening from sunset to sunrise hiking through and around an ancient quarry, Rano Raraku, where most of the stone heads were originally carved. Many had been abandoned in various stages of completion. The island's largest statue (about 70 feet in length) was here, lying on its back near completion and still attached to the mother volcano by a narrow strip of bedrock along its spine. All during the night we could see Bennett's Comet. Towards morning I set up my camera so that in the picture the comet would appear next to the nose of one of the stone heads. The sky was still dark enough for the comet to be seen even though the sun was coming up on the horizon, and the full-moon behind the camera highlighted some of the features of the statue. It was Palm Sunday, the week before Easter on Easter Island. That was an evening I'll never forget.

Another place of special mystery and beauty was Rano Kao—a volcanic crater on the southwestern corner of the island. An area along the top of Rano Kao has numerous stone carvings, pictographs, and caves associated with an old ritual honoring a species of birds that live on some small island pinnacles adjacent to the crater. The ritual was depicted in the 1994 movie *Rapa Nui* starring Jason Scott Lee. The floor of Rano Kao has a swampy environment with large reeds thought to have come to Easter via South America during early westward migrations. It has been suggested that the boats used in those migrations were made of similar reeds.

Bernardo and I decided to hike a few hundred feet down the inner wall of the crater to take a closer look at the reeds. It was more treacherous than we expected. There were constant small landslides as we went down. Once at the bottom we began hiking around in the swamp, looking at the different reeds and other vegetation. As we did this I was having a problem that didn't seem to be bothering Bernardo. Every once in awhile I would sink in up to my knees in muck. I was somewhat embarrassed because I obviously had no talent for walking on this stuff. Bernardo tried to explain his special technique—something about placing the ball of his foot down slowly and rolling gently backward onto his heel. I couldn't master this, and he was increasingly frustrated with my slowness. As we got into wetter areas, I sank in even further. Looking more closely at the water I noticed that it was bubbling upward through the muck. Even the normally buoyant Bernardo began to sink. Then a question flashed in my mind: *"Is this quicksand?"* From that point on I was scared—terrified might actually be a better word. I mentioned this possibility to Bernardo. He turned white and gulped. By now, I was getting in up to my waist and he was in up to his knees. Also, we had worked our way into the middle of the crater floor. We would now have a long way to go to get out. To keep from falling into the quicksand, we had to crawl, or swim through the wet areas that we couldn't avoid. Finally we made our way out of the swamp covered in mud, and soon began climbing up the inner wall to get to the top of the crater. As we climbed we were pleasantly surprised to find some wild grapes. We rested and ate, and our energy was restored. We began to think more rationally about our close call and Bernardo's alleged "swamp walking" skills. Upon further analysis it turned out that Bernardo weighed in at about 110 pounds, most of which were in his size 10 1/2 feet. I weighed in at 170 pounds with size 8 1/2 feet. In other words, not only did I have a bigger "cargo", but I also had smaller "boats". His skills and my shortcomings were merely validations of a simple law of physics (i.e., pressure = weight/area).

Reflecting on our adventure and Bernardo's alleged technique, we then had one of those laughing experiences that stay with you forever. We almost died choking on our grapes. I can't help but compare it to the scene in *Planes, Trains, and Automobiles* where John Candy and Steve Martin are sitting on Del's (i.e., John's) steamer trunk, reflecting upon their near death experiences from having gone the wrong way on a freeway. While they sat laughing at one another, their car behind them was being gutted by fire.

While Bernardo and I were having our laughs in Rano Kao, we had no car to catch on fire, but things were not going as expected back at the seismic station. During the morning inspection the next day, we found an unusual elevation of background noise on the records. The noise appeared to represent a rippling of the earth's surface under the station, and probably everywhere else on the island. Those rippling waves, just under the threshold of human perception, had periods from about 4 to 6 seconds between each successive peak. They were much larger than any noise waves either of us had ever seen before. Nor had we ever heard of such waves. We feared that something was wrong with our station. However, there were obvious recordings of earthquakes from a seismically active submarine ridge system located a few hundred miles to the west of the island. We concluded that since the station was recording earthquakes, the mystery noise was real and would, for the time being, have to remain unresolved. We were running out of time. Our flights were coming up—mine to Papeete the next day and Bernardo's to Santiago on the return flight from Tahiti. As we said our farewells on the following day, Antonio Haoa, a master carver, came by. I said "*hello*" and asked Bernardo to translate that I really liked his carvings and what a good time I had on Easter. At that point Antonio politely interrupted and said something that Bernardo had to translate. Antonio had stopped me when I said "*Easter*" to let me know that the real name of his homeland was not Easter. It was: "Te Pito o te Henua". A translation of this Polynesian phrase was "The Bellybutton of Everything", with the "Everything" interpretable as either the "Universe" or the "Earth". Considering elementary geography and the history of man, my initial reaction was that this perspective was totally inaccurate. His statement also confused me since it came from such a humble man. Later, I managed to sort this out on the flight back to Tahiti. After all, at one point in my life, I thought Vermilion was the center of the earth. Easter was thousands of miles from any significant landmass; the island had no direct contact with outsiders for many generations; and, in their eyes, the island was surrounded by a limitless expanse of water, skies, and stars. With these considerations "Bellybutton of the Universe" or "Bellybutton of the Earth" seemed to be totally appropriate terms. Little did I know at the time that the mystery of the noise would suggest that "Bellybutton" was a much more accurate term than any scientist could ever have imagined.

Where is Waikiki?

As I stepped off the plane at the old Honolulu terminal in August of 1963, I was greeted along with the rest of the passengers by a hula dance, a lei, and a kiss. My long planned escape into the world of endless summers was successful. I had been accepted as a graduate student in geophysics at the University of Hawaii. Of course, I'd miss seeing the Browns, Indians, St. Ignatius football games, and all my friends and relatives. But one has to live one's dreams. For the first time in my life, I could see a winter with no snow. I would not have to get up in the early morning of a 10 degrees below zero night to make my way from Westlake eastward across Cleveland to John Carroll University in Cleveland Heights. I would not have to scrape the ice off my windshield and hope that I could get the car up the driveway without shoveling too much snow. I would not have to spend up to four or five hours in a cold, icy, and dangerous commute along the Lakeshore Freeway and up Liberty Boulevard. I would not have to place a kerosene lantern next to my engine block and cover it with a blanket so that the car would start when I had to go to school or come back home. Instead, I would be swimming or surfing in Hawaii.

As I drove around Honolulu in my rented pink Jeep on that first day, I noticed more concrete than I expected. I saw Woolworth's but no grass shacks. Also, I couldn't find Waikiki! I kept driving around looking for a sign that said "Waikiki" or "Waikiki Beach". Furthermore, I couldn't even see a beach that resembled what I imagined Waikiki Beach had to look like. I learned later there were no signs; the view of the beach was blocked by countless hotels; and small sections of the beaches in the Waikiki district of Honolulu had different names or no names at all.

My initial disappointment of not finding the imagined Waikiki Beach would soon change as I began to explore the entire island. There were beaches elsewhere on Oahu that exceeded my most wild expectations of beauty. Before long I had an apartment with other students in Waikiki close to the University. I bought a surfboard and began learning to surf after school in the world's best known surfing playground, only five blocks from my apartment. As the winter approached and the surf came up on the North Shore, my roommates and some other students pooled our money together and rented a house near a surf-break called "Leftovers" between Waimea Bay and another break called Chun's. The fifteen of us had to pay about $10 each per month for the rent. With the "Beach Boys" music and the beach movies, surfing was in prime time during the 60's. I was lucky to be right there in the middle of it all.

At this point you may begin to suspect that I was a total goofoff. Actually, it was a case of play hard and work hard. I went to a rigorous private high school in Cleveland called St. Ignatius and majored in Physics at John Carroll University. I had almost no social life because of my studies. The atmosphere at the University of Hawaii's Institute of Geophysics was no different. Almost everyone there worked until late at night and on weekends. You could say that they were "A-type" personalities, but there was another factor—science at the Institute was fun!

In 1963 the Hawaii Institute of Geophysics (HIG) had only recently been established, and the man who was chosen to be its first director was probably the best person on the planet for that job—George P. Woollard. To me he always seemed much older than he really was. I could best describe him as a stern but benevolent grandfather type with an intense aura of wisdom and power. He had been the Director of the Geophysical and Polar Research Center at the University of Wisconsin and was soon to become the President of the American Geophysical Union. He had vision and knew with his efforts and Hawaii's potential, HIG could become an international leader in geophysical research. Prominent scientists from all over the world came to work for "Doc". However, his favorite people were his young graduate students. Although he would usually work them to exhaustion, Doc would always back them up in any way he could when they needed his help.

He often told me, as he did every other graduate student: "The world is your oyster. You can do anything you want." Another frequent statement was " everything is related". When he said that, I would always nod "yes", but would actually think he might be a little crazy. Certainly relationships could eventually be found between many, but not all, phenomena considered as separate and independent. I always wondered just how encompassing his *"everything"* was. He said it with such emphasis. I wish now that I would have had the nerve to ask him.

His attitude that "you can do anything" soon took hold. HIG began doing science all over the Pacific, in the most remote regions of the earth, from pole to pole, from the earth's inner core to the outer crust, through the ocean, into the atmosphere, to the moon, and to the limits of outer space. There was a collective belief that HIG could solve any problem related to the earth sciences and that we were proceeding to do just that. In other words we believed the Doc. However, my understanding of his *"everything is related"* would require years of work and future convergences of findings and mysteries.

Do It

A working group was being put together at HIG to look at hydroacoustic data from the Navy's Pacific Missile Range / Missile Impact Location System (PMR/MILS) sites. These were arrays of sensors scattered all over the Western Pacific to listen for the impacts of missiles launched from Vandenberg Air Force Base in California. The sensors were hydrophones in the ocean near Midway Island, Wake Island, Enewetak Atoll, and Oahu. When the missiles landed in the ocean, a small explosive charge was detonated, and the sound of the explosion was recorded by the hydrophones. Knowing the speed of sound in the ocean and the time at which the sound energy arrived at the different hydrophones, triangulation could be used to find the impact location and evaluate the missile's accuracy. HIG's job was not to determine missile accuracy—the Navy did that. It was to evaluate other signals in the hydrophone data and more accurately measure sound speed in the ocean. Many of the other signals were generated by earthquakes from around the Pacific, submarine volcanic eruptions often thousands of miles away, whales, shipping noises, or man-made underwater explosions.

Doc thought it would be good for me to work with this group because I had run a seismic station during the last three years of my undergraduate studies at John Carroll. I liked Doc's idea because hydrophones were a different way of looking at earthquakes, and the seismic energy would be traveling in an environment that was different from what was usually observed (i.e., under a purely oceanic crust as compared to being totally or partially under a continental crust). The opportunity for new discoveries was exciting to me. Within months of starting work, I had something—strange seismic phases unlike anything seen for continental travel paths. I showed them to Doc and suggested that if we were to fully understand these phases, someone would have to put conventional seismic stations on some of the remote islands in the Western Pacific. He agreed and said: "Do it!"

"Do it" was like a mini-lightening bolt. At first I was shocked. Then I thought he must be kidding. However, his stern expression didn't change. He was dead serious! Although I pretended to be pleased with his directive, I was very apprehensive. How would I "do it"? I was still wet behind the ears. Heck, I even had trouble packing my clothes to come to Hawaii. Seismic stations were big complex systems. To set them up on islands was no simple task, and was certainly too much for a young greenhorn grad student. He told me to write a proposal to the National Science Foundation and asked two other senior seismologist who were working on other projects to help me. The result was a rather bold proposal for a first year graduate student—installing seismic stations on Midway, Wake, and Marcus islands.

Not surprisingly, with Doc's reputation, my being known as one of his boys, and the interest of the National Science Foundation in helping HIG to become established, the proposal was funded. In the years that followed, with that initial grant and subsequent grants, I was able to become something of an expert on seismic and hydroacoustic data recorded on islands, in the ocean, and on the ocean floor. Upon completion of my Ph.D. in 1971, many of my colleagues were wondering where I would find a job after my current research grants ran out of money. There were no positions available for me at HIG. I wondered whether I would have to leave the land of endless summers. I decided to keep submitting research proposals, and to stay as long as I could.

Along with good science from the Midway, Wake, and Marcus stations, there were many great adventures and wonderful people on those islands. Midway is memorable for its wildlife, Wake for its lagoon and diving, and Marcus for it starry skies and access by way of Japan. All of these coral atolls were small islands with surface areas of only a few square miles.

Marcus Island

Marcus, about 1200 miles south east of Tokyo and more than 3000 miles west of Honolulu, was the smallest of the three. There was hardly enough land area for a runway for propeller driven aircraft. The most memorable experience of my trip to Marcus was a visit to the Japan Meteorological Agency (JMA) in Tokyo. At the time, ownership of Marcus was being transferred from the U.S. Department of Defense to the government of Japan. The only access to Marcus was by way of Japan. I was asked on very short notice to appear at JMA at a specific date and hour prior to my flight to Marcus. Although I didn't know why I had to appear, I showed up at JMA headquarters at the assigned hour. At the time of this trip, I must have been about 27 years old. I was brash, with no outward appearance of humility, and no knowledge of Japanese customs. Upon my arrival I was escorted to the top floor of the JMA building. Many older, well dressed men in suits and ties greeted me with nods. Although my aloha shirt and sneakers were of questionable taste, at least I was wearing long pants rather than my customary Bermuda shorts. To make things worse, I extended my hand to each of them. They obliged with unenthusiastic handshakes. We then moved into a special room for some tea. Following the lead of my host, I took off my shoes as I entered. I hadn't a clue as to what was happening.

Only one person addressed questions to me. He appeared to be the oldest and was obviously "the Boss". He never spoke English, although I suspected he could. The translated questions usually had their own built in answers. Our conversation went something like this. "So, you are going to Marcus?" . . . *"Yes, in a few days."* Then, after numerous non-translated discussions and exchanges between the other participants, the Boss would ask: "So, you are one of Professor Woollard's graduate students?" . . . "*Yes, I work for Doctor Woollard.*" Then, a few minutes later after more non-translated exchanges: "Do you like living in Hawaii?" . . . "*Yes, it's wonderful. The weather is nice all year long. It's not difficult to predict.*" I noticed that the Boss smiled slightly at my attempted humor—before the translation! Then after more exchanges: "Do you surf?" . . . "*Yes I do, in the summer in Waikiki and in the winter on the North Shore.*" And so it went.

Finally, after what seemed like hours, the Boss stood up and everyone followed. He wished me a very good trip. He said it was nice to meet me, and motioned toward the door. I said "*Aloha*" and nodded to all. As I put my shoes on, the sliding door closed behind me. Leaving the building, I checked my watch. I had been in there for about 40 of the most awkward, puzzling, and longest minutes of my life. Later, I would learn that Marcus was being turned over to JMA as a weather station and that JMA was one of the most powerful government agencies. Also, the Director of JMA was a cabinet level position in the Japanese government, and I had met him and other senior members of his agency. Wow! I hope I didn't screw-up too bad when I drank the tea!

Eventually I would come to understand that the meeting was primarily symbolic and ceremonial in nature—its purpose being the formal acknowledgment by the University of Hawaii of JMA's jurisdiction over Marcus and the formal receipt of JMA's endorsement of the University's research on Marcus. Many years later I would recount this story to a normally reserved, visiting colleague from Japan. He could hardly contain his laughter. After he composed himself, he told me that seismological research in Japan was under the direction of JMA; that he knew the Director; and that he had similar meetings in the same tea room. However, unlike the naive surf-bum American from Hawaii, he knew what was happening.

Wake Island

It should be noted that extended stays on small remote islands can drive otherwise normal people crazy. A prime example
of this occurred on Wake Island. Wake is about 2300 miles west
of Honolulu and 1500 miles east of Guam. Like many remote
islands, there was only one flight a week. I was on Wake trying to
get the new seismic station running, but wasn't having much
luck. I had to keep sending various components back to HIG's
electronics technicians for testing. This went on for so long that I
was running out of money and could no longer afford to stay in
Pan Am's "Quonset Hut Hotel". I found a mat and began sleep-
ing next to my instruments. People at HIG insisted that I leave
the island, but I was determined to stay until the system's problems were resolved. After several weeks of this, every component
had been checked and found to be functioning properly. The
system should have worked.

At that point I realized there was one final item which hadn't
been checked. It was a large transformer used to supply steady
voltages to the entire recording system. Such transformers were
very passive devices and were so uncomplicated as to never be
defective. Besides, it was new and had just been taken out of its
box at Wake. Whoops! That meant it wasn't tested with the rest of
the system in Honolulu. With that disturbing thought, I removed
the transformer from the system and plugged the individual components into the wall outlets throughout the room. Finally
everything worked. I was happy and furious at the same time. I
thought of the days and weeks of effort spent because of a defective transformer that was doing the opposite of what it was
supposed to do.

Considering all the trouble that the transformer had given
me, I could not simply dump it in a trash bin. Something more
symbolic and cathartic had to be done. I knew what! I'd throw it
in the ocean to give it a long and agonizing death in the salt
water. Better yet, I'd throw it over the edge of the reef where it

would also be pounded by ocean waves. That sounded terrific. I was feeling better already! Revenge was wonderful! The edge of the reef was less than a half mile from shore. The water depth to the edge of the reef varied from several inches to only a couple of feet. The sharks inside the reef were small. Waves were only crashing at the edge of the reef. The tide was coming up (but only a foot or two). The transformer weighed only 10 to 15 pounds. And, my flimsy rubber slippers were more than adequate for crossing the irregular surface of the coral reef. It would be a piece of cake. Yeah! Right! Sure!

Slipping around in my wet rubber slippers as I started to cross the reef, I began to think that revenge against an inanimate object might not be all that intelligent. The footing was bad, the water was deeper than I expected, the clumsy and heavy transformer was killing my back, and the tide was coming in faster than I expected. However, I was committed; and, it really would be wonderful to grab that thing by its cord and, like an Olympian in the hammer throw, swing that sucker over the reef. Unfortunately, as I proceeded, the water got deeper. It was already up to my chest in some places. By now the edge of the reef was fairly close, but the bottom was becoming more irregular and deeper. I realized I would have to dump the transformer here, but the water was so high I couldn't swing it around to give it a good toss.

Disappointed, I lifted the transformer over my head and shot-putted it out towards the edge as far as I could. Chest deep in water with bad footing, the distance amounted to all of about four feet. The instant before the release, my foot slipped and I felt a sharp pain. After my world record toss, I looked down at my heel. There was about a two-inch cut and blood was running out. Also, my slipper was broken and useless. Oh, what fun I would have! A half-mile hop across the reef with a rising tide. How much was I bleeding, and could a small shark track blood as well as a big shark? How small were those small sharks?

Clearly, I now fully realized that revenge against an inanimate object was not a good idea! Disgusted with myself, I ripped up my shirt, wrapped it around my foot, and made my way slowly and uneventfully back to shore. I've never before told the full story of my transformer adventure—only the good (*"I heaved the bugger over the reef"*), not the bad, (*"Boy was I stupid"*), nor the ugly (*"My brain was a mess and I really cut my foot"*). That afternoon, Wake had an operational seismic station; and, on the next flight, I went back to Honolulu with an unexplained limp.

Midway Island

Another example of the dangers of extended stays on small remote islands occurred on Midway at the northern end of the Hawaiian Islands about 1300 miles northwest of Honolulu. Midway has hundreds of thousands of birds, maybe millions. Most famous are the Laysan Albatross, also known as the "Gooney Bird". In spite of the name, they are no joke in the air. In that environment they are graceful, beautiful, and magnificent gliders. On the ground it's another story—more specifically their frequent crash and tumble landings. Because of this clumsiness, they are called "goonies" and are often thought to be stupid. Although clumsiness should not be considered as evidence of stupidity, I believe the goonies are, in fact, stupid. In support of this contention, I offer the following observations.

While waiting for the World War II vintage landing craft to take me across the lagoon from Eastern Island to Sand Island, I noticed a gooney in the water next to the pier that jutted out from the island into the inner lagoon. He was paddling into the wind, parallel to the pier, and towards the eight-to-ten foot bulkhead that separated the island from the lagoon. Paddling increased and merged into paddling with wings flapping. As he finally lifted off and pulled his feet under his belly, he noticed the bulkhead in front of him and realized that he wasn't lifting fast enough to clear it. With this realization his wings dropped, his webbed feet extended, and he hydroplaned gracefully on the surface of the water before slamming face-first into the bulkhead. He then proceeded to turn around and paddle back to the same point where the previous aborted take-off began. In the half-hour that I was waiting, he repeated this procedure seven times, each time becoming weaker and each time never moving further out into the lagoon to give him the extra range he needed. Fortunately, when the shuttle boat came, he went paddling off in another direction to get out of its way. Otherwise, he surely would have died trying.

The gooney was stupid, and I was getting there myself. After three or four weeks on the island, not all of the goonies looked the same to me. In fact, just from their appearance, I believed that I could tell, with reasonable certainty, whether or not a nesting gooney would bite when you attempted to rub its head. I often demonstrated this skill to my colleagues and new acquaintances. The frequent failures, of course, were painful and sometimes bloody. Obviously, three or four weeks on Midway were too much for me.

Island Travel

Yes, extended stays on small, remote islands can be dangerous to your mental and physical well being. Traveling to and from those places can also be similarly dangerous. In particular, I recall a flight that came into Wake from Hickam to leave off passengers and pick-up those who would be returning to Honolulu later that afternoon. After several changes in the departure time, it was obvious that something was seriously wrong with the plane. As we were trying to figure out what was happening, my colleague and I overheard one of the crewman say that they might have to wait until tomorrow because they didn't have the tools they needed. Being that it was Sunday, everything they might need, even if it were on Wake, was locked up.

Hey! We had a tool box right there in the lobby that we were going to hand carry on the plane. Although they probably needed something big or very special, we asked anyway: *"What do you guys need?"* They looked at one another, then at us, and said, somewhat sheepishly: "Have you got a hacksaw blade?" We couldn't believe our ears—a simple hacksaw blade! *"Sure, we've got the whole hacksaw, if you want."* . . . "Nah, it's in a tight place and we just need the blade." We gave it to them and they disappeared into the belly of the plane. A half-hour later they came back and said to everyone in the lobby: "Lets go." The engineer came over to us, returned the blade, and said: "Thanks!" He told us that either a spoiler or rudder (I forgot the term he used) was stuck and that the plane would crash if they took off and our hacksaw blade weren't used to cut away the obstruction. With that revelation, we weren't sure whether we really wanted to get on the plane. We did, but that's not the end of the story. As we rode on that C141 Starlifter back to Honolulu, we had the pleasure of watching a stream of hydraulic fluid run down the inner wall of the plane. Of course a crewman, who kept checking the leak throughout the flight, assured us that it was no big deal. Yeah! Right! Sure! Thanks a lot!

Global Warming and "Back to the Future"

A major question in the world today is how close humans are to the irrevocable destruction of an environment necessary for their continued survival. Byproducts of human activity have polluted the land, sea, and air. Human activities such as deforestation are also major contributors to environmental degradation. One of the most frightening scenarios being considered is that the pollution of the atmosphere with visible and invisible contaminants is already so bad that mankind's extinction is certain. If this supposition were proven to be true by no action other than hindsight, then all humans familiar with the supposition would, to varying degrees, share in the responsibility for the mass extinction of the most intelligent, creative, and loving species known to exist in the universe. Obviously, relying on hindsight is not an acceptable method for confirming or denying the supposition. For decades scientists have been conducting research on topics directly or indirectly related to all aspects of environmental degradation, including the topic of global warming. Unfortunately, the process of scientific inquiry does not always ensure an immediate or near term consensus on complex issues. The issue of global warming is no exception.

Some studies indicate that the earth is warming because of human activity, while other studies suggest that global warming is a natural perturbation that has happened throughout the earth's history. While the debate continues, many of the world's leading nations have enacted legislation to curb pollution and protect the environment. The cost of these measures, which in some instances may be unnecessary, is ultimately passed on to taxpayers, reducing in the short-term their quality of life. However, in the absence of solid scientific data, many of these hard governmental choices are understandable. Certainly there is a critical need for a method to accurately measure changes in worldwide temperatures to see if the earth is or is not warming, regardless of the cause.

Several years ago just such a method was suggested by one of the world's most well-known marine scientists, Walter Munk of the Scripps Institution of Oceanography. Dr. Munk and Andrew Forbes, an Australian oceanographer, proposed to use long-term variations of the sound speed in the oceans to measure global warming. They believed that this method would work because the variation of salinity, pressure, and temperature with depth in the ocean is known to produce a channel, or wave-guide, which traps sound energy. Furthermore the transmission of sound in the ocean is very efficient. Small sound sources can travel for thousands of miles with little loss of energy. They maintained that if global warming occurred, the warming of the ocean would produce measurable changes in the transmission rates of ocean sounds. Sound generators and worldwide transoceanic receptors would be needed to make these measurements. A multi-institutional coalition of hydroacousticians (i.e., experts in underwater sound transmission) proposed an experiment, and received funding to build and test a "boom box". This experiment became know as ATOC for Acoustic Thermometry of Ocean Climate. The subsequent tests indicated that the method would work.

As word of the experiment and its future objectives became public, environmentalist expressed concern for the effects that the boom boxes might have on marine life, especially whales. One of the problems with the boom box idea was that it would emit a relatively long-term unnatural tone that could drive marine organisms into abnormal behavioral patterns that could threaten their survival. Some of the creatures were endangered species. Putting it more simply, marine animals might be driven crazy in the same way that a dripping faucet drives people crazy. As with the dripping faucet, the problem with the boom box might be its unnaturalness and not its intensity. The concern of the environmentalists grew and a massive effort was undertaken to stop further experiments. Eventually, after many hearings and heated debates, the environmentalists succeeded in slowing the experiment and requiring studies on the effects of the boom box sounds on marine life. The results of these studies will determine whether the ATOC experiment receives continuing approvals.

Although it is not hard to imagine the frustrations of the ATOC scientists who had hoped to provide data that could be critical in identifying a new endangered species (i.e., Homo sapiens), I share the concerns of the environmentalists. I understand their skepticism of science and government. History is replete with examples of ill-conceived, government sponsored, and scientifically "justified" violations of the public trust. Yes, we do need good politicians and scientist. But to keep them that way, we also need intelligent, informed, rational, concerned, and involved citizens.

Getting back to the experiment, I believe that the controversy would never have occurred if the sound sources were nearly identical in their strength, frequency characteristics, and rates and times of occurrence to those sound sources that have been present in the oceans for millions of years. If scientist could generate such sources, time their origin, and time their arrivals at distant locations, measures could be made to determine the extent of global warming without seriously endangering other residents of our planet.

Actually, I believe that such sources are already available. Having personally examined hundreds of thousands of hours of recordings from hydrophones and island seismic stations, I, like many of my colleagues, know that the ocean is often filled with sounds from submarine volcanic eruptions originating along sub-duction zones and mid-ocean ridges. It is also known that the more violent episodes can be characterized by powerful short-term energy bursts that are somewhat similar in their frequency content, but greater in their intensity, to many man-made explo-sions which have been used for decades in geophysical studies of the oceanic crust, explorations for new oil fields in marine environments, military exercises, and in smaller scale studies of sound speeds in the world's oceans.

Some environmentalist should try to understand that the amount of energy needed for the transmission and detection of a signal over thousands of miles of ocean is far less than the amount released into the ocean by frequent submarine volcanic erup-tions or earthquakes. Small explosive charges of only one or two pounds of TNT, which are carefully timed and recorded, can be seen across entire oceans and could provide critical information on global warming. Yes, a few fish would be killed; but larger, communicating species such as whales or dolphins could be lo-cated with listening devices, and the detonations could be delayed until all such animals were out of the immediate area. I would ask those environmentalists who would find this unacceptable to conduct a reality check. They should ask themselves the follow-ing questions. "How many fish are sold in the world's markets each day?" "How many species other than man will become en-dangered, or extinct, if we do not quickly get a solid, quantitative, and unequivocal handle on present and future global warming?"

Too bad that a carefully timed experiment with small explo-sive charges wasn't conceived and implemented 30 years ago. Such an experiment could be repeated today to measure the extent of global warming during the past 30 years. Yep, it sure is too bad . . . Hey! Guess what? Although the world doesn't seem to know about it, just such an experiment was done more than 30 years ago! Furthermore, if prominent and politically well-con-nected environmentalists chose to do so, they could ensure that solid, quantitative, and unequivocal data on the extent of global warming over the past 30+ years could be made available to the world in a matter of days, or just a few weeks. More specifically, they would have to request that the Commander-in-Chief order the Navy to repeat an experiment which was completed in 1967 and published in the August 1969 issue of the *Journal of Geophysical Research.*

What's this all about and how does it relate to my story? I mentioned earlier that when I came to HIG, I was assigned to a group that looked at hydrophone recordings from the Navy's Missile Impact Location System (MILS). Well, if you're my age, some of your hard earned tax dollars were given to the Navy and used to fund a comprehensive, long-term study of sound speeds in the Pacific Ocean. It was a cooperative project employing top-secret sound surveillance systems and state of the art recording and analysis techniques of the University of Hawaii. In that experiment the Navy detonated ten explosions every month within an array of hydrophones near Midway Island for sixteen months. The times and locations of the detonations were determined with extreme precision by the Navy. Signals were recorded with similar precision on slow speed tape recordings of MILS hydrophones near Enewetak, Wake, and Oahu.

If the experiment were repeated today and if the actual warming of the earth over the past thirty years were even only a small percentage of what the ATOC group believes may have occurred, the newly measured values of sound speeds would clearly indicate those expected changes and provide the much needed unequivocal evidence of global warming. In fact, if the earth has warmed by only 0.05 degrees Fahrenheit during the past 30+ years, such an increase would easily be seen in a repeat of the 1967 experiment.

Now here's another connection to my story. Hydroacoustics began to emerge as a science during and after World War II as part of the U.S. Navy's effort to detect enemy submarines. One of the leaders in that research effort was Maurice Ewing of the Woods Hole Oceanographic Institution. During those embryonic experiments, the task of one of Dr. Ewing's unfortunate co-workers was to get into a rowboat far from everyone else; and, when told by radio to do so, drop charges of dynamite into the ocean. It was also suggested that the co-worker row quickly away from the area before the dynamite exploded because the boat might sink into the cavity of frothy water generated by the explosion. Surprisingly, the co-worker lived to recount his adventures. His name was George P. Woollard! More than fifty years later, Walter Munk and many other scientists were still talking about explosions in the ocean—not for man to extinguish submarines, but for man to see if he is extinguishing himself. Could this be one example of why Doc Woollard said: *"Everything is related."*

Also, you might wonder whether revelations of the 1967 experiment have been communicated elsewhere prior to their appearance here. They have—in the original publication in 1969 and more recently in *Eos*, a publication of the American Geophysical Union which is distributed world-wide to about 35,000 earth scientists ("Repeat of Thirty-Year-Old Experiment Could Prove or Disprove Global Warming" by D.A. Walker; 14 May 1996, pg. 191).

So far, I know of no response to, or interest expressed in, my call for a repeat of the 1967 experiment—a call for action which could provide the world with immediate, solid, quantitative, and unequivocal data on the extent of global warming over the past 30+ years. Emerging from this continuing silence are some disturbing questions. Is global warming politics, or is it science? Is science, science; or is science, politics? If an issue as important as global warming is more politics than science and if science is more politics than science, then our extinction may be inevitable.

Indeed, because a repetition of the 1967 experiment is such a "no-brainer", cynics could conclude that the experiment has already been repeated and that the results have remained a secret because they are not in the best interests of those who ordered the exercise.

Kaselehlia

On the island of Pohnpei (pronounced "pawn-pay") in Micronesia, Kaselehlia is the equivalent of Hawaii's "Aloha". With the mystery of the noise at Easter Island still fresh in my mind, I landed on Pohnpei (then it was called Ponape) in April of 1972, just two years after my Easter Island adventures. I arrived with several hundred pounds of seismic instrumentation, an accumulated wealth of sixty-five dollars, and a wife who shouldn't have been on the plane because she was more than six months pregnant. By then I had some ideas about the noise, but wasn't certain. Hopefully, the data from Pohnpei would help to resolve that mystery.

Pohnpei is about 3000 miles south west of Honolulu and 1000 miles south east of Guam. In the early 70's it could have been compared in many respects to Hawaii just after the turn of the century. This was a far greater step back in time than I expected. It was especially hard on my wife with our first baby due in a few short weeks, but nothing really bothered me. I had my seismic station and the island looked like it would be the scene of many great adventures. Although it was small (probably $1/3$ the size of Oahu), it had beautiful mountains, rivers, waterfalls, and a distant fringing reef.

It took a few days to get the station running properly, and to find transportation and a place to stay. No sooner had we gotten into a routine, than a typhoon approached the island. The safest place for us to hide in our flimsy house was under a heavy dining room table. We lived there for two days as the typhoon pounded the island. Every once in awhile, we'd look outside (half the house was nothing more than a screened-in porch) and see parts of roofs and trees flying by. The sound was terrifying—like hundreds of jets circling overhead. My part-Hawaiian, totally pregnant wife surely must have wondered what she had gotten herself into, marrying this crazy scientist who brought her to such a primitive island, and who seemed to care more about his seismic station than her and their soon-to-be born daughter.

Unlike everyone else, especially my wife, I thought of how lucky I was to be there, and to have arrived in time to set up my equipment and record the typhoon. I was interested in how much energy would be pumped into the earth by the storm and its effect on the seismic records. As the typhoon approached, the noise levels on the recorders became higher and higher, even though the winds were not yet felt on the island. Apparently the storm was not only moving horizontally over the earth's surface but vertical fluctuations in its power (like a giant plunger) were transmitted through the water to the ocean floor and were generating waves in the crust of the earth. Although these waves remained below the threshold of human perception, they were becoming so large on the instruments that they threatened to break the recording pens by swinging them all the way into the side of the recorder's housing. Just before the typhoon was about to hit, I had to shut the system down. Fortunately, my instruments were located in the island's typhoon-proof communications center that had its own emergency power. As soon as the storm weakened, and while my wife remained under the table wondering even more about my sanity, I worked my way back up to the station and fired up the system to track the typhoon as it continued heading northwestward away from Pohnpei and towards Japan. I had already noted earlier, during its approach, that the intensity of the background noise appeared to be comparable to, or even larger than, the mystery noise that I had seen at Easter. It seemed, therefore, that the mystery noise at Easter could have been caused by a nearby hurricane or storm. This possibility had been considered before coming to Pohnpei, but satellite photos showed no indications of storms near Easter. Now it seemed that all of the data would have to be looked at more carefully.

The Pohnpei station had the standard north-south, east-west, and vertical components of long-and short-period seismometers. What all this meant was that within a few minutes of a large earthquake occurring almost anywhere in the world, my station could record microscopic pulses of energy from the earthquake that had traveled through the earth at speeds of many thousands of miles an hour. I could tell just how much the earth had moved at Pohnpei, as well as the direction and period of the ground motion (i.e., the time between each successive oscillation of the earth). With such information gathered by seismic stations all around the world, seismologists have made encyclopedic contributions to the understanding of the content, history, and dynamics of our earth. Also there have been enormous offshoots of earthquake seismology. As the world's appetite for oil increased, seismologists developed smaller instruments and began to generate artificial earthquakes with dynamite blasts, so as to study in more detail, the structure of the earth's shallow crust. Before long most of the world's oil, on land and below the seas, would be discovered through seismic exploration with devices manufactured by such pioneering oil exploration companies as Texas Instruments. Furthermore, their continuing push for miniaturization led to the commercial development of the transistor—launching the era of semiconductors, which was the birth of today's personal computers and the earth's information highway, and without which space exploration, as we know it, would not exist. Indeed, all aspects of our lives (be it medical research, communications, transportation, or "you name it") have been affected by this technology. Doc, is this once again an example of *"everything is related"*?

I had no aspirations of spawning another industrial revolution with data from the Pohnpei station. I just wanted to make a modest contribution to the earth sciences by acquiring a better understanding of the structure of the earth under the Western Pacific Ocean. Then again, at one time some seismologists only wanted to make smaller, simpler seismic stations; and look what happened! Maybe I could do more than learn about the earth under the Western Pacific. Maybe the typhoon and the mystery noise could tell me something else about the earth. Unfortunately, the Easter Island records and the satellite data that I needed were back at the University, and I still had several more months of data to acquire for my studies of the earth's structure under the Western Pacific. I would have to wait long after my daughter was born on the 4th of July in 1972. I would not return to the University until the fall of 1973.

The greater size and diversity of Pohnpei compared to Midway, Wake, and Marcus permitted sufficient levels of sanity to be maintained for many months. Also, my incidents of "loosing it" on Wake and Midway (especially Wake) may have helped me to recognize and avoid similar incidents on Pohnpei. Hiking and diving were the primary source of entertainment and adventure. Although we usually snorkeled in shallow waters close to the island, on one especially calm day, we decided to scuba dive in the deep ocean along the outer edge of the fringing reef a few miles from the island. My fellow adventurers were a couple of newly arrived contract workers from California who did not have much experience in tropical waters. As a result they relied on my judgement and seemed more comfortable in the ocean because of my presence.

I remember being amazed at how small the island appeared as we anchored our boat on the outer reef. The diving was great. There were many beautiful fish and corals. Throughout the dive I was always worried about sharks since they could be much bigger in those deeper waters. After a couple of hours, we worked ourselves back to the boat. As we were getting in and taking off our gear, I said: *"I'm sure glad we didn't see any sharks, or I would have been out of there!"* . . . "What! You didn't see the sharks?" . . . *"No."* . . . "But they were there, almost all the time." . . . *"How big were they?"* . . . "Four, five, maybe six feet." . . . *"I never saw them. Where were they?"* . . . "They kept cruising along the outside of us. We could barely see them cause they were so far away, but they were always following us." . . . *"Geez guys, I wish you'd have signaled or something."* . . . "We thought you saw them and it was OK." . . . *"Guys, I'm nearsighted and don't have prescription glass in my mask. I can't see that well in the water!"*

After the chilling revelations of our mutual misunderstandings, we had little to say to one another on our trip back to the harbor. Needless to say, those guys never went diving with me again. On subsequent dives I found new partners, advised them of my handicap, and confined myself to the safer environment of the shallower waters closer to the island.

One of the most pleasant and humble individuals that I met during my stay was a local native who maintained the grounds around the island's communication center where I had my recording instruments. His name was Yoshda. He was fascinated with the recordings of distant earthquakes. When I first showed him the system, I told him to stand very still. Then I turned the magnification all the way up and told him to move his body slowly up and down without taking his feet off the ground. As he did this, the pen made a peak-to-peak tracing of several inches on the paper that followed his every move. To him it was total magic. Yoshda would often bring his friends and relatives. Although I'd have to hook up a special seismometer in the room and put on a new sheet of paper so as not to mess up the data already recorded, the reactions of amazement and laughter defy description, and were well worth my efforts for their entertainment value.

On the Monday after my diving adventure, I told Yoshda about the sharks my friends had seen on the outer reef. He said we didn't have to worry. He fished for them all the time and fed them to his family. They tasted good. In fact, he had caught a shark along another portion of the outer reef on the day of our dive. I asked him what kind of boat he had and how he caught sharks. What he told me was a technique that I suspect not everyone will rush out and try. His boat was an outrigger canoe that he made from a breadfruit tree. His equipment consisted of a harpoon, a couple hundred feet of rope, and several large fishing floats about the size of basketballs. He would paddle to the edge of the reef, jump into the water, spear some small fish, mash them up pretty good, spread them around, and wait there—in the water! Pretty soon, out of the darkness of the distant waters, the sharks would emerge. He would harpoon whichever one he liked, swim back to his canoe, and let out the line with the floats attached. He couldn't tie the rope to his little boat because either his boat would be swamped or the rope would break. [Remember what happened to the transom of the Orca, and to Quint in *Jaws*!]. So, he would just let the rope and floats go, and spend many minutes, and sometimes hours, following the floats around the ocean as they occasionally popped-up. Eventually the shark would get tired and die. Sometimes the floats would go down and he would never see them again. He liked to eat big sharks! The one he harpooned that Saturday was about 10 feet long. He had to track him for almost four hours. The floats were under the surface for so long and were so far away when they came up that he was afraid he would loose his equipment. Yoshda was less than 5 feet tall. Quint and Matt were total wimps compared to Yoshda!

Later, on another day, the outer reef would provide some additional anxiety. The purpose of that day's activity was to gather shells—no transformers, no scuba gear, no snorkels, and no visual impairments. Just a simple walk on the surface of the reef at low tide in tennis shoes with my glasses on to see any small unfriendly creatures that might be in the shallow water, generally not much more than a foot deep. I had gloves and a long screwdriver to pry and probe, without having to worry about being bit by an eel. I had even taken the precaution of putting a strap on my glasses so they wouldn't slip off. All aspects of the adventure were carefully planned, conservative, and safe. Yeah! Right! Sure! Tell me more!

We (i.e., myself, an architect, a school administrator, and their wives—mine was home taking care of the baby) went out to a northern section of the reef, not too far from my earlier diving adventure, in a twelve foot Zodiac with a 25 horsepower Johnson engine. We anchored in the shallow water and proceeded with our explorations. While everyone else headed east of where the Zodiac was anchored, I figured there would be better hunting alone, to the west of the Zodiac. I was moving rapidly along the reef, and in no time was a few hundred feet from the boat. As I glanced back at my friends, I noticed some dark clouds on the horizon behind them, where just a few minutes before there had been sunny skies. The wind also started to pick-up, as well as the height of the choppy, shallow water waves.

I had seen storms like this before in Ohio on Lake Erie. I knew we could be in for some big trouble. I shouted to get everyone's attention and pointed to the black clouds behind them. As we all quickly headed towards the boat, the skies darkened, it started to rain, and the wind and waves got even stronger. Deep concern turned to horror, when we saw that the wind loosened our anchor rope and the Zodiac was blowing away, across the reef toward the deeper water of the inner lagoon. I thought that my friends would be able to get the boat because they were much closer to it than I was. However, the Zodiac was moving faster than they could run in the rapidly rising waters of the reef.

As they saw the boat moving farther and farther away, they must have thought that they might die. There was no visible end to the storm. In a few hours it would be dark and the tide would be coming in. The reef was a few miles from the island, which earlier wasn't much more than a speck on the horizon. Now, it couldn't even be seen. If the weather cleared in time, it would be impossible for anyone on the island to see a few heads bobbing up and down out on the reef. There were no planes or rescue helicopters to hope for. Even if the men could swim to shore, would they leave their wives stranded on the reef? Would they know in which direction to swim? Would the waves and currents allow them to reach the shore, or would they drown quickly or be swept away never to be seen again?

Because I had gone alone (sometimes it pays to be a loner) to the west of the Zodiac, my position on the curving reef was more perpendicular to their line of pursuit and the heading of the boat. I watched as the gap between them and the boat continued to increase, and I was petrified. When they soon realized that they weren't going to catch the Zodiac, they all began yelling: "Dan, get the boat! . . . Dan, get the boat!" By now the strong winds had blown the boat into the deeper waters of the inner lagoon, and it was moving fast. I was the only one who had a chance to get the boat, but I would have to act fast. I fixed my eyes on an imaginary point several hundred feet ahead of the boat and began swimming in that direction. I'm sorry, but I won't be modest on this point. I was a great swimmer in Vermilion as a kid, and an even better swimmer in Hawaii with all my surfing. I have no doubt that in those few minutes I set a world record that will not be broken for a very, very long time. I was a finely tuned machine, propelled by instinct and adrenaline. During the swim I can only recall the sensation of being almost above the water. Finally, just as the boat was about to pass me by, I reached out and grabbed a rope hanging along its side. A few more seconds and it would have been too late. I pulled myself on board and collapsed on the floor.

My friends began yelling for me to start the engine and come back for them; but I couldn't move because of exhaustion. They kept yelling. Perhaps they thought I had passed out, and they were trying to wake me up. I could hardly hear them with the wind and rain and distance involved. When I finally got up, I noticed that my gloves were no longer on my hands, and my tennis shoes were no longer on my feet. Indeed, I must have been flying through the water! Fortunately I was still wearing my swimsuit and my strapped-on glasses. The engine started after a few anxious pulls, and I headed back out to my friends. By then they were up to their chests in water, and the waves were getting bigger. They climbed in and we headed towards where we thought the island had been. The storm intensified and waves began crashing over the Zodiac. The engine was hit by a wave and it stopped running. Other waves started to fill the boat.

The only good news was that Zodiacs don't sink even when they're totally swamped. Clearly we were not in a nice place, but it was better than where we were before, on the reef and up the proverbial creek without a paddle, or a boat! We could only hope that the storm would end quickly and that we could get the motor started again. We were so messed up though, that we wouldn't know where to steer the boat even if the engine did start. The storm just had to end soon—that's all! It would be dark in a few hours. Meanwhile, back on the island, the storm damaged some roofs and knocked down some big branches. It shook our house and banged its doors. Francine knew we were out there and was aware of the fact that people were frequently lost at sea throughout Micronesia. She thought that she might not see her husband or friends again.

As the storm began to clear, the suspense was overwhelming. If we saw the island and could get the motor started, we might make it. If we didn't see the island, we would probably die a long and horrible death. As the darkness moved on and the late afternoon sun began to appear, the island emerged slowly out of the clouds. Screams of relief and joy rang out! *"Amen, Alleluia, and Praise the Lord!"* Land was only a few hundred yards away. We could row if we had to, but the engine started anyhow. We had been blown around to the northwest side of the island. Once we got to that remote and uninhabited shore, we could not bring ourselves to head back to town by way of the now quiet ocean. Instead we worked our way along a narrow channel in a mangrove swamp that cut through the corner of the island. We arrived in town near dark.

When I got home I couldn't bring myself to tell Francine

everything that had happened; nor to this day does she fully understand all of the other possible outcomes of that adventure. Although it's not in the *Guinness Book of World Records*, I can give you the names of four people who will swear that they were eyewitnesses to what may have been the fastest swim in the history of the human race. That night I decided I had enough data. It was time to pack up the instruments and head back to Hawaii . . . Kaselehlia!

The Revolution and SALT

While I was having my adventures on islands across the Pacific, my colleagues in geophysics were having their own great adventures all over the earth. A revolution in thinking and discovery was sweeping the field. Critical elements of that revolution were the completion of a worldwide standardized network of seismic stations and the development of various instruments for studying the crust and uppermost mantle of the world's oceans.

The new worldwide seismic network provided comprehensive information on the frequency, size, location, and depth of earthquakes all over the earth. Composite maps of earthquake locations revealed a continuum of epicenters winding in irregular patterns all over the surface of the earth. Many of the epicenters were either along irregular lines passing through the middle of the earth's oceans or along the margins of continents. Plots of depths revealed that earthquakes in the oceans were never much more than a few miles deep, while earthquakes along continental margins had depths ranging from a few miles to hundreds of miles. Also, along continental margins, the quakes were generally shallower closer to the ocean and deeper inland under the continent. With the new seismic data, the earth looked like an orange that had been clumsily pealed by hand and then reassembled. In other words, its surface was a bunch of plates. The edges of some continents were either climbing over and depressing the edges of oceanic plates or the oceanic plates were diving under and lifting up the edges of continental plates, or both of these processes were occurring to some degree.

New marine geophysical instruments provided for more detailed studies of various physical properties of those margins. Thus were born or confirmed, such concepts as sea-floor spreading and subduction. The "Theory of Plate Tectonics" soon became an accepted fact. As with so many other great discoveries, the explanations were so simple and comprehensive that it was difficult to understand why the theory wasn't thought of earlier.

Well, the truth of the matter is, it was! In 1910, Alfred Wegener, a German meteorologist, proposed that at one time South America and Africa were joined together, along with all of the other continents. He believed that over the course of time, this super continent was broken apart and scattered across the globe. He provided, even by today's standards, solid scientific data to support his conclusions. For his visionary efforts, he was widely ridiculed by prominent geologists of the time, and he had to endure that scorn for the remainder of his career. After his death in 1930, Wegener and his ideas were soon forgotten.

Was it presumptuous for a meteorologist to tell geologists about their own field? It probably seemed so to some of the geologists, and that may have been a significant factor in the rejection of his ideas. Although we would like to think otherwise, "intelligent" does not necessarily mean "smart" (i.e., considerate, thoughtful, imaginative, dedicated, open-minded, honest, and humble). In addition, unrestrained egos often prevent the most "intelligent" from ever becoming "smart".

Unfortunately, there weren't enough smart and powerful geologists around to help Wegener. If Professor Wegener were here today, I would tell him that he was a great visionary scientist and that the other guys weren't even in his league. I would like to ask him if he shared Doc's view that: *"Everything is related"*. After all, he was a meteorologist and he brought continents together. The plate tectonics revolution proved that he was right, and united what previously had been thought to be many unrelated observations, phenomena, and processes. As a witness to the revolution in the late 60's and early 70's, my understanding of Doc's "*everything*" continued to grow.

Also during the 60's and 70's, a less exciting but potentially more important topic which began to emerge was that of arms proliferation. The State Department was becoming increasingly concerned with the spread of nuclear weapons. That shared concern with a number of other nuclear powers resulted in a long series of international negotiations which became known as the "SALT" conferences—SALT being an acronym for Strategic Arms Limitations Talks. The objective of SALT was to limit the size and number of tests by existing nuclear powers, and to prevent the spread of nuclear weapons to other nations. A critical issue was whether seismologists could detect very small underground explosions and whether they could distinguish man-made explosions from naturally occurring earthquakes.

Now back at HIG after my most recent island adventures on Pohnpei, I needed to analyze the hundreds of earthquake phases that were recorded there and to compare the Pohnpei typhoon noise and the mystery noise at Easter. A more important concern, however, was the fact that my funding was running out. If my summers were to remain endless, I'd better line up some new sources of long-term support. One possibility was to become involved in SALT. I thought my chances might be good because back in 1963 when I began working for Doc on the MILS data, I noticed that the hydrophones were recording underground nuclear explosions from Russia and Nevada. Late in 1969, I did a preliminary study of these and other nuclear explosions. At the time, I had no ulterior motive in working on these data. It was just an interesting thing to do. Now, a few years later, it seemed that the research might be important to some people in Washington, D.C.

The recording of nuclear explosions on hydrophones was a bit of a surprise. Hydrophones were high frequency instruments. Based on observations from most continental seismic stations, nuclear explosions recorded at great distances were not supposed to have that much energy at those high frequencies. Could the ocean floor be a quieter environment than continents at those higher frequencies—so quiet that a broader range of signals from nuclear explosions could be observed and analyzed? That was a question I asked in my proposals to the agencies sponsoring nuclear test detection research. Furthermore, I noted that the Wake hydrophones were roughly equidistant from test sites in Nevada, Tahiti, China, and Eastern Russia. Comparisons of recordings in those differing geological environments could be very useful in determining the detection and discrimination capabilities of seismic and hydrophone stations. My proposals were funded. The endless summer in Hawaii would continue.

During this time I also gathered up the recordings from Easter and Pohnpei. I wanted to compare the noise produced by the Pohnpei typhoon to the noise at Easter. That mystery had been bothering me for too long already. I had earlier determined that the mystery noise was not a continuous feature of the Easter Island recordings. When it was present, it's intensity and duration were quite variable. Sometimes the noise would continue for 2 or 3 weeks, at other times it might last for a few hours. On closer inspection of the Pohnpei records, I found that the largest noise levels from the typhoon did not have quite the same periods (i.e., the time between consecutive peaks) as the noise at Easter. Furthermore, based on the instrument magnification, the motion of the ground at Easter for its longer period noise was much greater than the ground motion for the shorter period typhoon noise on Pohnpei. These findings and my earlier search of satellite data indicated that the noise at Easter was not generated by storms. I still had a suspicion of what its cause might be, but I needed more data from Easter Island. With the political turmoil in Chile during and after the presidency of Salvadore Allende, communications with Chile and Easter ceased and the station was abandoned. The data I needed, and thought I might never see, would eventually come from a very unlikely source—strange sightings off the coast of Japan thousands of miles from Easter.

Kaitoku and the Mystery Cloud

Presumably, the Navy was getting a little better with the accuracy of their missiles from Vandenberg. They were now able to put them into the lagoon of Kwajalein Atoll, more than 4,000 miles away, rather than scatter them all over the Western Pacific. [Were they ever really that bad, or was the big ocean-wide PMR/MILS system also used to find those boats that never come up for air?] As a result, the hydrophones at Midway, Wake, Enewetak, and Oahu were no longer needed. If they were being used for submarine detection, more sophisticated devices and methods had made them obsolete. Eventually, they were all to be abandoned or cannibalized.

With its extensive and diverse distributions of hydrophone arrays and sub-arrays, the Wake site was the crown jewel of the entire network. The abandonment and possible destruction of this multi-million dollar facility would be a great loss for seismology and ocean acoustics. Fortunately, our nuclear test detection research required the reactivation of the Wake array with new slow-speed tape recorders and computer-controlled signal processing devices. Instead of remaining abandoned, the Wake system was turned over to the University of Hawaii only a few short months prior to its scheduled demolition.

In the years of operation that followed, we were able to demonstrate that the deep oceans are extremely quiet sites for recording nuclear explosions and that nuclear explosions were different from earthquakes in the relative strength of their signals at different periods. Also, with our sophisticated recording and analysis techniques, we were able to more carefully look at some of the discoveries made with the earthquake data from Pohnpei. We were able to publish a number of interesting research papers on our findings.

Unexpected surprises in the data recorded by the new Wake system were several episodes of submarine volcanic activity, mostly from the Japan to Guam portion of the circum-Pacific subduction zone. I recalled seeing these kinds of episodes in the PMR/MILS data back in 1963. My colleagues in that group had published several papers on submarine volcanism. A comparison of the new data with the old would be interesting. [I was glad that I had salvaged tons of old recordings from the trash bin when the PMR/MILS study group disbanded many years earlier.] The contracting agencies, the Air Force Office of Scientific Research and the Arms Control and Disarmament Agency, didn't mind some peripheral investigations of these eruptions, but our primary focus was, of course, underground nuclear explosions. We mostly noted the times of the volcanic activity, and made some cursory comparisons with the "trash bin data", never really studying the eruptions in any detail.

Then, one day shortly after 9 April 1984, I received a call from D.C. It was an Air Force Colonel whose name I have since forgotten. He asked that we look for evidence in our data of submarine volcanic activity off the coast of Northern Japan. It seems that crew members and passengers on military and commercial jetliners en route from Tokyo to Anchorage saw what appeared to be an unusually large, rapidly expanding, perfectly symmetrical, optical anomaly high in the moon-lit skies east of Honshu. A volcanic eruption was one of the possibilities the Air Force was investigating, and the Colonel was aware of our ability to detect submarine volcanic eruptions. The sightings had been covered extensively by the Alaska news media with interviews of pilots, crewmen, and passengers of the different planes. The so-called cloud was estimated to be about 200 miles in diameter.

The Colonel wanted some quick answers. "How soon can you look at the data?" . . . "*Well, we just received a tape from Wake the other day, so it will be about another week before we get the tape that covers the day you're interested in.*" . . . "Can't you have them send it today?" . . . "*No. There's only one flight a week. If you want it quicker, you'll have to send a special plane to get it.*" . . . "Damn! OK. Let me know what you find out as soon as you can." . . . "*Sure thing. It'll probably be in seven to ten days.*" . . . "All right. Talk to you later. Bye."

The task ahead would require much more work than I expected. Although it eventually resulted in a publication in *Science*, the effort seemed professionally to be a waste of time. I didn't feel too bad though, because many of my colleagues told me that it was really a neat paper. Also, a junior editor of *Science* told me that the Editor-in-Chief found it especially interesting. I wrote it like a science mystery. To do a good job, I had to cross some disciplines and had to re-interview some of the witnesses.

I hate to think of what would have happened if the phenomenon occurred today. There would be the standard media frenzy. We would probably be seeing it on *Nightline*, *Hard Copy*, *Inside Edition*, and *Extra*. Talk shows would be swarming all over the witnesses. There would be UFO supporters and debunkers, conspiracy theorists, scientists, and shrinks. And, it would probably be a story line for a future episode of the *X-Files*. All the more so, because to this date no one has solved the mystery. [If any "Ruskies" are reading this book, and you know anything about this, please let me know. If you tell me and come to Hawaii, I'll take you surfing and be your tour guide!]

Getting back to the data, there was no apparent submarine volcanic activity in the Aleutians or Kurils on that day or in the weeks immediately before or after the sightings. However, there was a powerful, long-lasting episode of submarine volcanism between Japan and Guam near Kaitoku Seamount. In fact the episode reached its peak intensity around the time of the sightings. Too bad Kaitoku was about 900 miles away from the location of the mystery cloud. If it reached the ocean's surface in a violent eruption, it could have put a somewhat symmetrical cloud into the atmosphere. But what mechanism could have moved such a cloud northward while maintaining its integrity, and what mechanisms would cause it to suddenly expand? We gave it a good try, but couldn't come up with any acceptable answers.

I still wonder why the Air Force requested that the University of Hawaii look at data from the Wake hydrophones. Didn't they have access to all of the "secret" arrays of military hydrophones from Guam northward to the Kamchatka Peninsula used to monitor Russian submarine movements? Did they actually want us to conclude that sea-floor volcanic activity could have been the cause of the mystery cloud?

"We have a round ball cloud . . . looks like a nuclear explosion, only there was no fireball and there was no lighting . . . but the cloud was there very definitely. We're at altitude 33,000 feet . . . the cloud continues to explode, like a great big cloud . . . easy to see it. The moon is behind it . . . it expanded very rapidly." [Captain, Japan Air Lines flight 036.]

"Cockaroaching" and Good-Bye

Hollywood

University researchers generally have what is called either "hard money" or "soft money" positions. Hard money positions are internally supported and usually lead to tenure (i.e., job security). Soft money positions, often called "Post-Docs", are externally supported with no long-term security. Usually a Post-Doc's income disappears when his or her research grant runs out of money. For this reason, Post-Docs have an extra incentive to explore ideas that could secure future funding. However, most research grants do not provide for such unauthorized explorations.

Hawaii has a few bugs with which the population is continually waging war. One is the notorious cockroach. Worldwide, the cockroach is known to sneak out of its hiding place, take something off the counter or floor, and return quickly to its nest. Thus, over generations of warfare in Hawaii, the term "cockaroach" has evolved. To "cockaroach" something is to secretly borrow an insignificant object as a joke from a friend, either until its absence is discovered or it is no longer needed by the "cockaroacher". Post-Doc's are "cockaroaches" because they have to "cockaroach" time from existing research projects to explore for new sources of funding.

As an individual on soft money for 32 years, I was an expert "cockaroach". Nearly all of my research support came from look-ing at things during prior research projects that I was not supposed to look at. One of my unsuccessful "cockaroaching" expeditions involved the global triggering of earthquakes. It was unsuccess-ful in the sense that it never resulted in any research grants or contracts. However, it was successful in that it led to some inter-esting research papers, and battles with editors and reviewers.

My global triggering adventures began as part of my effort to solve the mystery of the noise at Easter. Having scanned that data for the noise study, I decided I had better, at long last, look at the earthquakes on those records. The primary earthquake regions were the submarine ridge system which ran in a north-south direction west of Easter and the subduc-tion zone far to the east of the island. The submarine ridge system is the East Pacific Rise (EPR), and the subduction zone is the west coast of South America—with its most promi-nent features being the Peru-Chile Trench and the Andes. The EPR stretches from California to nearly Antarctica where it splits and heads off towards the Indian Ocean.

In my Easter Island data, there were a number of small earth-quakes recorded from the EPR and a few large earthquakes from South America. The largest and most destructive earthquake re-corded was the large Peruvian quake of 31 May 1970. It had a magnitude of 7.6 and killed more than 67,000 people.

It would be nice if seismologist could predict earthquakes. According to plate tectonic theory either sea-floor spreading causes subduction, subduction causes sea-floor spreading, or both processes might be occurring. Perhaps some temporal correlation might exist between the activity along the East Pacific Rise and South America even though these plate edges were more than 2,000 miles apart. To some this would imply the ridiculous suggestion that a single large earthquake, or a number of smaller earthquakes, along one edge would be powerful enough to move the entire oceanic plate, producing subsequent earthquakes on adjacent edges. I agree that such a process would be unlikely. However, that was no reason not to look for correlations. Maybe something else might be going on. Perhaps the earthquake waves themselves, which travel in a matter of minutes across oceans, could trigger regions that were close to failure. Perhaps there were other kinds of waves we did not know about. Perhaps the whole earth acted as a system responding to some unknown internal, or external, fluctuating stress field that triggered earthquakes either regionally or globally. Who knew for sure? It was worth a look. Besides, it wouldn't take that long.

To make a long story short, a correlation was found between earthquake activity on the rising and sinking edges of the plate. The most intense episode of seismicity on the EPR occurred at about the same time as the large Peruvian earthquake. The study was expanded to other related rising and sinking plate edges. Additional suggestions of correlations were found. Technical reports and research papers were published.

However, on one occasion it was obvious that the editor was making me spin my wheels and that he would never let my paper be published. I wondered whether he was blocking my paper because of its scary implications or whether he was trying to protect me. Regardless of his reason, and in spite of my frustration and disappointment, I think the rejection was in my own best interests. Had I published that paper and the logical, obvious follow-up that I had "waiting in the wings," I would have been ridiculed and would not have received the research funding needed to defend myself. Most likely my career as a research scientist would have come to an end. You see the paper that I had "in the wings" suggested that underground nuclear testing in Nevada had triggered earthquakes in California . . . "Whoops! Sorry about that San Fernando!" . . . "Sorry about those thyroid cancers, too!"

Anyhow the large tests have stopped, so what's the point. Well, I guess the point is the same one that was made earlier. Good science is essential to our survival; intelligent scientists can screw-up; smart scientists are less likely to screw-up; and an informed, intelligent, and concerned citizenry can help keep the screw-ups to a minimum.

I gave up trying to get those papers published. Instead, I submitted a very simple report to *Science* which showed plots of the yearly seismic energy released worldwide by large earthquakes and the energy released by ridge system (i.e., spreading center) earthquakes. Similarities in the trends of the plots suggest that subduction zone earthquakes are triggered by ridge earthquakes, ridge earthquakes are triggered by subduction zone earthquakes, both are triggered by some other force, or some or all of the above.

Initially, the editor refused to publish the paper because a British reviewer implied that if I had used more accurate British data, rather than inferior United States data, my correlations would not have been as good. Although it took a bit of work (again of the "cockaroach" variety!), I recomputed the plots with the British data. The correlations weren't worse, they were better! The paper was published. Pretty good stuff that British data—I've used it ever since!

More recently, I published an updated version of the *Science* paper and added an analysis of large rare earthquakes. There are many regions of the earth which only rarely have large earthquakes. For example, one such region is Yap, an island about 1400 miles west of Pohnpei. It has had only three large earthquakes thus far in this century. They occurred in 1911 (a 7.6), 1912 (a 7.4), and 1914 (a 7.5). Another region is Nankaido, Japan which had large earthquakes only in 1944 (a 7.8), 1946 (a 8.0), and 1948 (a 7.1). The study suggests that regions all across the earth that rarely have large earthquakes have them most frequently during episodes of intense worldwide seismic activity—the most recent episode began in about 1989 and included unusual sequences of earthquakes in the Japan and California regions.

Finally, getting back to Nevada, I've often wondered whether those guys there couldn't have distributed a sequence of bombs in time and space so that only certain portions of California would fall into the ocean? . . . Just kidding!

The Earth's Bellybutton

Most of the seismic events recorded at Easter Island from the East Pacific Rise (EPR) were not to be found in earthquake listings of the U.S. National Earthquake Information Service (NEIS) or in the more comprehensive listings of Great Britain's International Seismological Centre (ISC). This was not surprising because, aside from the Easter Island station, which was not part of the worldwide network, the nearest seismic stations were in Peru and Chile, more than 2,000 miles away. Such distant stations could not "see" any of the very small, shallow earthquakes from the ridge system to the west of Easter. The large number of these otherwise unknown earthquakes was surprising. They were identifiable as earthquakes by their differing seismic phases.

The so-called P-phase ("P" for primary) of earthquake seismology is nothing more than the compressional energy that travels through all media (e.g., air, water, rock) when it is generated by a disturbance. Talking and hearing are good examples of P-phase generation, transmission, and reception. The vibrations of the voice box are the earthquakes, the propagation of sound through the air is the transmission, and the eardrums are the seismometers. P-phase transmission is a molecule by molecule (or, possibly, atom by atom) transfer of energy—like a line of billiard balls touching one another, where the last ball goes flying away from the rest of the balls when the first ball in the line-up is struck by the cue ball. The so-called S-phase (S for secondary) is also a molecule by molecule transfer of energy of a different sort. Imagine if you could, a string of single molecules lined up end to end. Tie one end to a fence, move back, and pull the string tight. Now, quickly shake the string vertically up and down. A pulse of waves, like ocean waves with peaks and troughs, will visibly deform the string and move towards the fence. Because P-phases and S-phases require a molecule by molecule transfer of energy, they can be thought of as travelling within "the body" of the material. For this reason they are called body phases, or body waves. [My apologies if I may be loosing some of you. Hang in there! I'm almost done, and this is important!]

Along with body waves, another type of seismic energy transfer is through so-called surface waves. Imagine that the crust (or, if you wish, one of the layers of rock that makes up the crust) is represented by a heavy carpet. You tie the ends of the carpet to the fence at several places along its edge, step back, and pull it tight so that the whole carpet is parallel to the ground. [If you've imagined too big of a carpet, either think smaller or use more people to hold it up.] Now shake it vertically up and down as you did earlier with the string of molecules for the S-phase generation. A pulse of waves will move across the entire carpet towards the fence.

In an earthquake the entire crust, and sometimes deeper layers, are rippled by surface waves while at the same time P and S waves are moving internally through the entire "body" of the earth. These motions spread out at different speeds in all directions from the source of the earthquake with variable intensities, depending on the geology and nature of the earthquake. Finally, a special type of energy encountered in ocean environments is the so-called T-phase (T for tertiary). It is nothing more than a P-phase in the ocean that is generated when the energy from an earthquake leaves the solid earth on the ocean floor and enters the water.

In the recordings at Easter, all of the above mentioned phases, each with its own distinctive arrival times and characteristics, could be identified for the larger reported earthquakes from the

EPR. By comparing phases of unreported earthquakes to the phases of reported earthquakes, most were identified as being from smaller earthquakes in the same area of the EPR. Further investigations revealed that reported earthquakes from the EPR often occurred during swarms of unreported earthquakes. Also present in such swarms were intense high-frequency bursts of energy. I had seen similar energy bursts back in 1963 and in the Kaitoku study. These were T-phases from shallow submarine volcanic eruptions.

Obviously, the reported earthquakes, the swarms of unreported earthquakes, and the swarms of isolated T-phases were all evidence of an intense episode of sea-floor spreading, or "the birthing" of young crust. This added new meaning to Antonio's admonition regarding the name of his home. Could he or his ancestors have actually believed in a literal meaning of the term "Bellybutton of the Earth". Furthermore, if any place on the face of the earth could be called "The Earth's Bellybutton", the plate tectonics revolution had now proven that it was near Easter Island. The ridge system there was found to be the most rapidly spreading on earth. It was so fast that in some places it couldn't keep up with itself, and small mini-plates were forming. In 1992 and 1993, hundreds of volcanoes were found along this ridge system. Also, many hundreds of thousands of years earlier, Easter had been right on the axis of this most rapidly spreading ridge center. Easter itself was formed by the energy (lava) travelling through umbilical columns of conduits from its Mother (the mantle), which gives birth to all of the earth's crust. Easter Island is now merely a remnant of that birthing process (i.e., a bellybutton).

Wow! I almost wonder now whether that was really an admonition from Antonio, or whether it was God talking through him and having some fun with me!

Great Convergences

Studies of the reported and unreported earthquakes along the East Pacific Rise (EPR) tweaked some dormant cells in my brain. They barely had enough energy to ask a long forgotten question: "What about that noise? . . . "*What noise? Oh. Yeah. That's right, the mystery noise! But I gave up on that.*" . . . "Look again."

Not one to ignore tweaking brain cells, I began flipping through the seismograms looking at the noise. Now the margins of my Easter seismograms were all marked up with notes from the previous study of the reported and unreported East Pacific Rise earthquakes. "*That's strange*", I thought. "*The episodes of mystery noise seemed to occur on those records where the margins are marked up the most.*" In other words, the noise seemed to be related to episodes of ridge system earthquakes and possible T-phase volcanism. "*But what was causing the noise?*" The answer soon became obvious. The strongest phases for many of the reported EPR earthquakes were the surface waves. These 4 to 6 second period oscillations lasted for a few tens of seconds before they dropped down into the background noise. For some of the unreported earthquakes, all that could be seen above the background noise were several cycles of their surface waves. During these episodes of seismic activity, the mystery noise was often the lower level background noise, and it had the same 4 to 6 second period as the earthquake surface waves. In other words, the mystery noise was produced by a continuum of small earthquakes.

The mystery, at long last, was solved. Like the sounds of the other earthquake phases, the sounds of the mystery noise were the sounds of labor and delivery—the sounds of Mother Earth and baby—the sounds of new crust forming by sea-floor spreading.

The plate tectonics revolution had focused a great deal of attention on ridge systems. Many oceanographers of differing disciplines wanted to look directly at what was happening in the ocean and on the ocean floor during episodes of spreading. However, to know when such an episode was occurring, one would need seismic and acoustic data to monitor the level of spreading activity. The old PMR/MILS data salvaged earlier from the trash bin, the Kaitoku data, and the Easter Island data were used to confirm the potential of sea-floor spreading event detection through the use of hydrophones, and to help convince the Navy to allow the use of previously classified hydrophone arrays. With such instruments, episodes of sea-floor spreading off the coast of Oregon and Washington have since been detected as they occurred, permitting on-site studies while spreading was in progress.

Event detection in the oceans would prove later to be of interest to another group of scientists. You may have heard of the Comprehensive Test Ban Treaty, or CTBT. After SALT (i.e., the Strategic Arms Limitations Talks) came CTBT. Almost every nation on earth was interested in signing this treaty which would outlaw testing of nuclear weapons in the air, in the solid earth, and in the oceans. For the treaty to work, and before anyone would sign it, some fundamental conditions had to be met. Everyone had to have access to raw data so that they could independently search for violations; and the data gathered had to be of sufficient quality to detect violations in every environment. An environment of great concern to many of the potential signatories was the ocean. Their questions were: "Could man-made explosions in the ocean be distinguished from earthquake noises or submarine volcanic eruptions; and could instruments in the ocean detect very small explosions?" I got a phone call, and it was back to the Easter, Kaitoku, and "trash bin" data.

The proof was there. Hydrophones would do the job. The treaty was signed, reducing, at least temporarily, the long endured threat of nuclear annihilation.

The approval in 1996 of the Comprehensive Test Ban Treaty by 158 nations in the U.N General Assembly was a significant milestone in one of the most difficult and complex negotiations in history. One can only have the greatest respect for those involved in what must have been, at times, a grueling, frustrating, and depressing process. Generally unrecognized by the public is the broad spectrum of scientific disciplines and decades of collective research required for this achievement. It was wonderful to have been a small part of that effort. The greatest reward for me was the knowledge that the Wake hydrophone array would become part of the CTBT network; and that its data would be available for monitoring and research for many years to come.

Moving on to another convergence, I noted earlier that the mystery noise was, in effect, the sea-floor spreading, that sea-floor spreading occurred episodically, and that the ridge next to Easter was the earth's most rapidly spreading ridge system. Those findings prompted the following thoughts. Throughout history we have often observed the effects of land-based volcanic eruptions on the earth's climate. Yet the world's ocean ridge systems contain many thousand of active underwater volcanoes, and we hardly even know that they exist. We are impressed by the power of large earthquakes, which is often translated from a magnitude into its equivalent in terms of hundreds or thousands of atomic bombs. Also, we are impressed by the power involved in the uplifting of continents to produce great mountains. But where does the power of earthquakes and mountain building come from? It comes from sea-floor spreading. The oceanic ridge systems must be an enormous energy source, and we may know very little about the effect of these energy sources on our environment because they are hidden under a blanket of oceans.

There wasn't much more that I could do with the Easter data except to appreciate the new perspective of ridge system power that my recordings had given me. It was time to move on to other studies. However, that move didn't last long. As though attached by a mental bungee cord, I was soon snapped back to "The Earth's Bellybutton", to the silent stone heads that appeared as if they possessed answers to all the secrets of the universe.

My mental journey was triggered by an article in a science magazine that discussed whether an El Niño was predicted anytime soon. As usual, some experts said "yes" and others said "no". I read the article because I was now a little more interested in things atmospheric, having looked at satellite photos of the Easter Island area and having studied upper atmospheric air flow over Kaitoku. The article went on to explain that a primary indicator of El Niños was relative changes in the atmospheric pressure over Darwin, Australia and Easter Island! "Goosebumps or chicken skin"—call it what you like. I was transported back in a nanosecond. No sooner had the word "Easter" entered my brain, than I pictured the seismograms with the mystery noise, the reported and unreported earthquakes, and the T-phases. I simultaneously said to myself: *"Wow! I wonder if the El Niños and episodes of sea-floor spreading happened at the same time? Was the physics right? Would the pressure variation at Easter require an energy input or an energy output?"*

The first thing I had to do was acquire a better understanding of El Niños. I read many articles and research papers. The critical elements that I found were the following. (1) The term El Niño translates to "The Child". Historically, Peruvian fisherman would observe occasional episodes of unusually warm coastal water around the Christmas season. For some unknown reason they chose to associate this warm water, which killed or drove the fish away, with Christ's birthday, and called it "The Child". Why they didn't call it something else, like "bad December water" is beyond me. (2) Another term for the phenomenon is ENSO for El Niño/Southern Oscillation Index. (3) The Southern Oscillation Index (SOI) is a measure of the difference in atmospheric pressure at Easter Island and the pressure at Darwin, Australia compared to long-term averages in those values. Now, you ask: "What does the atmospheric pressure at Darwin, or for that matter at Easter, have to do with the migrations of fish off the coast of South America?" Well, to my surprise I learned that (4) one of the most dominant high pressure cells in the earth's atmosphere is generally centered over the area of Easter Island. Why it's there has to do with many factors—the primary ones being the relative distribution of continents and oceans, the rotation of the earth, and solar energy. (5) Moving westward from this high pressure cell, the next cell to be encountered is a low pressure cell centered roughly over northern Australia and Indonesia. (6) These two cells feed into one another and control atmospheric and ocean surface circulation throughout the South Pacific. (7) Normally, this massive engine works fine; but every few years or so, it gets some bad gas or something. It sputters and doesn't work too well. The atmosphere becomes disturbed, ocean currents weaken, water warms, fish are confused (they go elsewhere and/or they die), droughts occur in regions that are normally wet, and torrential rains occur in regions that are normally dry. In fact, most atmospheric scientists believe that some El Niños have severe global effects. Phenomena often attributed to El Niños include: changes in the strength and direction of the jet stream; record high temperatures in China, Alaska, and the northern and southern U.S.; above average rainfall in the Southwest, the Gulf Coast, and the Southeast; droughts in Indonesia, Australia, and Africa; flooding in parts of South America, Asia, and Europe; outbreaks of cholera, hantavirus, dengue fever, and plague; marine mammal strandings; fewer hurricanes in the Atlantic; and more hurricanes and typhoons in the Pacific. (8) All of this leads to the fact that El Niños are important, and it would be nice to know when they are coming. Careful planning and long-term emergency preparedness could lesson the economic losses and personal tragedies. (9) No one knew what caused them or if there was a distinct cause. (10) When El Niños occurred, the SOI was unusually low; and when the SOI was unusually low, an El Niño was in progress.

In other words, there was a one-to-one correlation; and that's why El Niños also became known as "ENSO episodes". (11) Analysis of the Index during times of El Niños indicated that sustained pressure changes required to produce an El Niño were extremely small. (12) El Niños generally last for 1 or 2 years. (13) Their recurrence intervals are irregular and generally range from about 2 to 6 years.

At this point anyone would have to consider how remarkable it was that the earth's greatest crust building engine and the earth's greatest atmospheric engine were in the same place. In view of all the other *"everything"* relationships that had evolved throughout my career, I had to wonder now whether the influence of the "The Earth's Bellybutton" also extended into the earth's atmosphere. The problem in looking for possible correlations between El Niños and intense episodes of sea-floor spreading near Easter Island was that I only had data from the Easter station for a couple of years. Therefore, some of the most accurate indicators of sea-floor spreading (i.e., swarms of unreported earthquakes, T-phases from submarine volcanoes, and anomalous noise levels at Easter) were unavailable for a long-term study. All that I could use were the origin times, epicenters, and magnitudes of earthquakes which were reported from the EPR in National Earthquake Information Service (NEIS) or International Seismological Centre (ISC) listings.

Unfortunately, the existing Easter Island data indicated that reported earthquakes were only the "tip of the iceberg" in terms of spreading activity. For example, if there were a few reported earthquakes from the spreading center in a month, there could be many more unreported earthquakes, all kinds of T-phases, and many days of "mystery noise". Although the available data did indicate that reported earthquakes were associated with increased levels of these other sea-floor spreading indicators, there was still the possibility that increased spreading might not always be associated with increased numbers of reported earthquakes. This possibility might be less likely if only the most intense episodes of reported earthquakes were compared to the times of El Niños. Continuing with the iceberg analogy, the maximum ice that can be seen (i.e., the most intense episodes of reported earthquakes), the most likely it is that even greater amounts are under the surface (i.e., the most intense episodes of sea-floor spreading).

The only reliable data for the region was that acquired since the completion of the World-Wide Network of Standard Seismographs in 1964. Data compiled from these stations by the NEIS and ISC from 1964 through 1987 were used. Plots of the times of occurrence and magnitudes of these earthquakes were compared to the five El Niños during that time interval. The correlations are striking. The largest earthquakes and/or the highest numbers of reported earthquakes occurred during the times of El Niños. Although the long-term monthly number of reported earthquakes from the region averaged only about 2 per month, months with eight or more reported earthquakes were found in El Niño years. In some instances as many as 16 earthquakes in a month were reported with epicenters distributed over hundreds of miles of the ridge system . . . *"Had the silent figures and their Mother finally revealed their greatest secrets to me?"*

In the paper that was subsequently published in *Eos*, some suggestions were offered as to how the energy from the sea-floor might be transferred to the ocean's surface. It was noted that the pressure change required to start an El Niño was only about 2 or 3 parts in 1000; and that such a change was easily achievable if only a minute portion of the sea-floor spreading energy could get to the surface by direct or indirect means. Also, the physics was right. We all know that hot air rises because hot air is lighter than cold air. Atmospheric pressure is nothing more than a measure of the weight per unit area of the atmosphere above us. [Yes, the air is made up of molecules; they push down on us; and they have weight]. That pressure is 14.7 pounds per square inch on the surface of the earth at sea level. In metric terms it's 1000 millibars, and an El Niño can start with sustained changes in pressure of only 2 or 3 millibars. That's where the above mentioned 2 or 3 parts per 1000 came from.

The possible worldwide effects of such small changes (i.e., the disruptions to worldwide weather patterns caused by El Niños) confirms what the average person already knows. The atmosphere is usually in a state of unstable equilibrium and small changes can have significant consequences. Furthermore, once equilibrium is disturbed, it may take a long time before the atmosphere returns to normal. With my research appearing in the most widely distributed publication in the earth sciences, I waited with great expectations for some feedback and interest from my colleagues. I waited, and waited, and waited, and kept waiting!

The Second Silence

If nothing else, I'm a persistent and generally positive person. I tried not to think of the silence. The only feedback I received was from reviews of my unsuccessful proposals for further research that were usually given to oceanographers or atmospheric scientists. To many of them the ideas of this seismologist to explain an oceanic and atmospheric phenomenon were preposterous. Anyhow, they probably believed that their extensive ocean-wide surveys of temperatures and currents, as well as supercomputer modeling of the atmosphere would soon find El Niño's trigger. My hopes to reactivate the Easter station and to install a hydrophone array to seismically and hydroacoustically monitor sea-floor spreading, and possibly predict El Niños, were unrealistic. I would have to wait for another El Niño. It came in 1991. For this new El Niño, I extended the earlier study from 1987 through 1992. Once again the correlations were there, and once again I included new suggestions of possible mechanisms for the transfer of energy from the sea-floor to the ocean's surface. Those new suggestions were based on recent oceanographic research papers. Once again I hoped my colleagues would be excited about the idea and would field test my suggestions or try some of their own. Again, I waited, and waited, and continued to wait. Now the only feedback was generally negative "shoot from hip" comments by earth scientists to inquiries from newspaper reporters or editors of magazines. It was obvious from their statements that many of those respondents had not read my research papers. This was not the constructive forum I had hoped for.

I wondered how many more El Niños it would take, and when the next one would happen. It didn't take long. On 5 September 1996, seismic data suggested that a new El Niño would soon begin. Subsequently, a flattening of the Southern Oscillation Index was reported for September, and the Index began to plummet. Several months later, reports of a new El Niño began to appear on an almost weekly basis in the popular press, in science newsmagazines, and on television. It was reported that climate experts were "taken by surprise".

The media frenzy continued throughout 1997 and early 1998, as long as there were sensational weather related disasters and reported Pacific equatorial surface waters that were warmer than normal. By the end of 1998, I had completed my third paper suggesting a link between EPR seismicity and El Niños. In a few months my updated research on this topic was once again published in *Eos*.

Now with seven El Niños and 36 years of earthquake data from the East Pacific Rise west of Easter Island, let me summarize the apparent correlations. In the horizontal rows of information which follow: the top row of numbers indicates the decade; the second row indicates the year from 1964 through 1999; the third row indicates those years in which El Niños occurred (if less than ½ of the year was an El Niño year, it is not indicated); the fourth row indicates whether large earthquakes were observed for any month in that year; the fifth row indicates whether an unusually large number of earthquakes were observed for any month in that year; and, to help in the analysis, the bottom row again indicates the El Niño years.

DECADE	60's						70's										80's										90's									
YEAR	4	5	6	7	8	9	0	1	2	3	4	5	6	7	8	9	0	1	2	3	4	5	6	7	8	9	0	1	2	3	4	5	6	7	8	9
EL NINO	X	X								X				X	X				X	X			X	X					X	X	X	X		X		
HIGH ENERGY	X	X							X										X	X		X	X								X					
MANY QUAKES	X		X										X						X				X							X		X	X	X		
EL NINO	X	X								X				X	X				X	X			X	X					X	X	X	X		X		

Scanning vertically up and down the columns, it is obvious that large earthquakes and large numbers of earthquakes occur during, or near to, the years in which El Niños have been observed. Returning to the iceberg analogy, we don't know whether the bulk of the iceberg (i.e., the unreported phenomena associated with sea-floor spreading) is directly below the observable ice (i.e., the intense episodes of reported earthquakes), is to its left, or is to its right (i.e., occurs in the same year, in the previous year, or in the following year).

Do the apparent correlations of El Niños and earthquakes necessarily imply that the phenomena are related? No, they don't. However, the irregularity of the recurrence rates suggests that it is more likely that they are related. Therefore, objective scientists are duty-bound to examine possible mechanisms that could link intensive episodes of sea-floor spreading near Easter Island to El Niños.

Hypothetically, possible linkages could occur in two directions. Atmospheric pressure changes could trigger sea-floor spreading or sea-floor spreading could trigger atmospheric pressure changes. Considering the first option, some of the largest short-term pressure changes observed at Easter in non-El Niño years could not be found to be associated with EPR earthquakes. Furthermore, although such triggering of regions close to failure might occasionally be possible in far less active areas of the earth, it is difficult to imagine how fluctuating atmospheric forces of no more than 2 or 3 ounces could dominate the enormously greater force fluctuations continually present below sea-floor spreading centers.

Considering the possibility that sea-floor spreading might trigger El Niños, we note the following facts. (1) Sea-floor spreading near Easter Island is putting large amounts of energy into the water at the floor of the ocean. (2) The "tip of the iceberg" concept observed at Easter has since been substantiated along other portions of ocean ridge systems. (3) The heart of the high-pressure cell in the southeastern Pacific is approximately over the spreading center. (4) Lows in the Index can be produced by raising the pressure at Darwin or lowering the pressure at Easter. (5) The dominant factor during El Niños is the lowering of pressure at Easter and not the raising of pressure at Darwin. (6) Heating of the atmosphere will lower its pressure. (7) Plankton blooms in surface waters can increase the opacity of the water and increase heating of the air. (8) Deep-ocean plankton feed on nutrients supplied by sea-floor spreading. (9) Some of the body parts from deep-ocean plankton float to the surface and are a potential food source for surface water plankton.

These facts provide suggestions as to how intense episodes of sea-floor spreading could heat the atmosphere and trigger El Niños. There are other possibilities including disruptions of intermediate-depth ocean currents, direct transfers of large super-heated and super-buoyant plumes of water to the ocean's surface, and lower levels of atmospheric heating that might cause a migration of the high-pressure cell. Other, as yet undiscovered, mechanisms are possible. I wonder whether the earth science community will continue to disregard the apparent correlations of El Niños and spreading center seismicity. If so, how many more correlations with future El Niños will be needed before the issue is given the careful, objective scientific scrutiny it deserves.

In 1986 a preeminent meteorologist, Colin Ramage, said in his article on "El Niños" in *Scientific American* that: "the failure to predict El Niño underlines the current lack of understanding of how the anomaly develops." Fourteen years later atmospheric scientists and oceanographers may be no closer to a solution, even though hundreds of millions of dollars have been spent during this time to find the triggering mechanism.

It's interesting to note that the terms "Southern Oscillation Index" and "ENSO episode" are rarely referred to in media discussions of El Niño, in spite of the fact that oceanographers and atmospheric scientist are well aware of the unquestionable, constant linkage of El Niños to the Index. However, many of the scientist interviewed by the media leave the public with the perception that an El Niño is nothing more than a massive pool of unusually warm water extending from east to west across the equatorial Pacific. Perhaps they feel that the general public is either incapable of understanding, or is not interested in, a more comprehensive explanation. Perhaps they have decided that those beautiful color pictures of the temperature anomaly across the equatorial Pacific are sufficiently convienent and dramatic representations of the phenomenon. Perhaps they are right!

However, if truth and understanding are important, the equatorial band of warm water is far from being the whole story. The normally relentlessly inquisitive media seem to have sheepishly accepted this limited, simple-minded explanation. Few journalists have pursued the question of: "What causes the warm water?" Indeed, the warm water is merely one of the more dramatic elements in a long chain of events resulting from an unusual lowering of the pressure in a massive atmospheric pressure cell covering much of the southeastern Pacific. Again, atmospheric scientist and oceanographers know this, and that's why they often refer to the phenomena as "ENSO episodes" rather than "El Niños".

Thus, a simple explanation for the cause of the warm water, consistent with accepted doctrines of oceanographers and atmospheric scientists, is that a big water pump (the southeast Pacific atmospheric high pressure cell) helps to push water from east to west across the equator. The fuel for this pump is cold air; and if that fuel is warmed, the pump will not work very well, or it may not work at all.

Another aspect of El Niños that requires further explanation is its usual observation by South American fishermen around the time of Christmas. Regardless of when an ENSO episode occurs, South American coastal water temperatures will most dramatically exceed the tolerance of marine organisms when ENSO warming pushes temperatures beyond the already high values experienced during the southern hemisphere's summer—"around the time of Christmas". When this happens the fish will leave or die, and fishermen whose livelihoods depend on their presence, will be the first to know. They will once again note the apparent cruel connection of their economic depression to the season of "The Child's" birth.

Finally, as you may have gathered from most of my story, I've had good reason throughout my career to be positive and optimistic about science. Now, however, I'm worried about science. I wonder whether individuals with integrity, dedication, intelligence, and vision are still able to do their job. Big science seems to be getting bigger, while small science seems to be getting smaller. Does this encourage a "go along" to "get along" attitude that is the antitheses of creative thinking? It is also surprising to me that after years of "failure", new approaches—even those that cut across scientific disciplines—would not be welcomed. Perhaps I'm still too young and impatient; colleagues will look more seriously at my suggestions or other explanations; great leaders will once again emerge and say, like Doc would, "Damit, we'd better take a look at that!"; and the silences (i.e., the 1967 ocean sound-speed experiment, and the possible sea-floor spreading /El Niño connection) will be no more. I hope so!

Sometimes

Sometimes things don't go as planned. Sometimes that's good. Many of the greatest scientific discoveries were made by accident and by those on a path not frequently taken. Careful planning, rigid intelligence, and highly focused research has often been proven to be a less successful strategy for discovery than random explorations of curious departures from mainstream thinking, by fresh and open minds, by voracious appetites for all kinds of good data, and by luck.

Many years ago I saved some eloquent explanations of this often neglected fact and taped the clippings to the side of a bookcase in my office. As my research progressed the clippings soon were covered by an adjacent bookcase and became lost from sight and memory. With the office cleaning and customary redistribution of furniture to colleagues upon my retirement from the University of Hawaii in 1995, the clippings reappeared with chilling relevance to my career. Although the examples cited in the following *Science* editorial and letter to the editor are from the field of medical research, these perspectives similarly reflect career experiences across a broad spectrum of scientific disciplines.

"A Sense of the History of Discovery. Far too many physicians and scientists, as well as laymen, look upon the history of discovery as an entertaining pastime, a tiresome academic exercise, or merely the record of egotistical aspirations. But it is essential to understand the successes and failures of others. Take three well-known examples of success.

Alexander Fleming made probably the most important medical discovery of the century, through a combination of almost ludicrous circumstances. He was favored by fortune to have a mycologist working on *Penicillium notatum* on the floor above his laboratory. Fleming's laboratory was primitive, and outside air circulated in a way that only a rugged Scotsman would tolerate. Furthermore, he failed to wash his petri dishes before he went on vacation. Using common sense, Fleming followed up on an observation that many had made before him. He looked at the petri dishes when he returned from vacation and did something about what he saw. Had Fleming been an active committee attendee, with plenty of dishwashers and technicians, it is unlikely he would ever have noted the clear zone around the bacterial colony, because a dutiful technician would long since have destroyed the evidence even if it had been allowed to develop.

A second example—the discovery of liver extract for the treatment of pernicious anemia resulted from work by Whipple, Minot, and Murphy. The pathologist Whipple was trying to determine the hematopoietic efficiency of diets. The clinician Minot was broadly enough read not only to know of Whipple's basic work, but to see its applicability. This led to its clinical use by Murphy, and the award of three Nobel prizes.

The discovery of insulin by Banting and Best tells another story. Even though it had been conclusively shown that the pancreas was crucially involved in diabetes, none of the major research centers did much about it. It took a young unknown surgeon with avid curiosity and an equally eager young medical student to do what should long since have been done. Note that government grants and major planning committees representing consumers, economists, sociologists, et al. played no part.

Today many investigators fail to appreciate that knowledge of the history of discovery is vital if wise policies are to be generated. Instead the young are being taught, by example, that status and celebrity seeking promote their careers. Serving on important committees, belonging to the right societies, associating with the right people, and seeking discreet but maximum public exposure pay handsomely—if this is the way of life that is sought.

Government, with the power of money at its disposal, plays a critical part. The power-oriented among us find it easy to accept the hegemony of government over practice and research, forgetting that a sense of the history of discovery is almost wholly absent in those making policy. Are such important decisions to be left in the hands of those so innocent?

I am certain that the conduct of research will ultimately triumph over the business and politics of discovery. I am less certain about the practice of medicine, to the politically powerful a much larger plum. It is especially difficult to disentangle those practicing medicine in the broad sense from those who would control it but without experience to guide them.

Perhaps the most we have the right to expect is that some people may heed a twinge of conscience. But the issue is now clear: if greatness is a goal, it will take great thinking and consummate honesty to achieve it. History has shown us and formulated the guidelines.—Irving H. Page, former Director, Research Division, Cleveland Clinic, Cleveland, Ohio 44106." [*Science 186*, 1161, 1974.]

"Obvious Question. I was much interested in Irving H. Page's editorial "A sense of the history of discovery", especially the description of the ludicrous circumstances under which Fleming discovered penicillin. I think I can add a further ludicrous note to the discovery of penicillin.

When I was an undergraduate in medical school and taking a bacteriology course (in 1914), we learned how to grow bacteria on agar plates. One day my plate had a number of black spots on it surrounded by clear halos. I asked the instructor what those clear halos were containing a black spot in the center; I don't recall his exact words, but the tenor of his response was, "Those are molds: you were careless in your technique and you got your plate contaminated by molds. You must be more careful."

I am sure that bit of knowledge was not his alone. The other instructors and the professor of bacteriology must have know also of the black spots surrounded by clear zones. There must have been hundreds of bacteriologists around the world at that time who had seen this same thing. Incredibly, it seems that the perfectly obvious question, "If something diffuses out from a colony of molds which will prevent bacterial growth in culture, might this also prevent infection in man?" seems not to have occurred to any of them. Why didn't that so very obvious question occur to me? Instead, I went back to my place thoroughly chastened, having been chided before the whole class for carelessness in technique. Before the day was over, all my classmates knew that molds destroyed bacterial growth. They were all reasonably intelligent; why didn't the question occur to one of them?

If a reasonably intelligent and curious high school student had wandered in to visit the laboratory, he, being thoroughly disinterested, might very well have asked, after the situation was explained to him, "Is that what you use in sick people to kill bacteria?"

It has always seemed to me that this was a prime example of how extremely obtuse even intelligent people may be. After all, the only reason we were studying bacteriology was to learn how to control infectious diseases!—A. C. Hilding, Research Laboratory, St. Luke's Hospital, Duluth, Minnesota 55805." [*Science 187*, 703, 1975.]

A Smoother Pebble

Among the estimated 35,000 earth scientist who subscribe to *Eos*, the journal in which my research on El Niños was published in 1988, 1995, and 1999, it seemed most logical that suggested mechanisms, in addition to those that I had proposed, would come from oceanographers and atmospheric scientists. Little did I suspect back in 1988 that my first suggestion would come from a space center scientist ten years later.

Physics 101

His mother abandoned him as an infant. He never knew his father. He wasn't very good as a farmer or as a student. He was absent-minded and graduated from college without distinction. Some professors complained about his lack of knowledge of mathematics. Nonetheless, he eventually became regarded as the most original and important theorist in the history of science. Although Isaac Newton was born on Christmas Day more than 350 years ago, Albert Einstein said that his own work would not have been possible without Newton's discoveries which "are even today still guiding our thinking in physics."

Perhaps Newton's greatest contribution was the Universal Law of Gravitation. This law states that every object in the universe attracts every other object in the universe with a force that is related to the amount of matter in each of the objects and the distance between them. [A force is a push or pull tending to cause motion, and the amount of matter in an object is called its mass.] As it turns out, Newton determined that gravitational force is directly related to the mass of an object, so that if one of the masses is doubled, the force will be doubled, or if one of the masses is cut in half, the force will be cut in half. Regarding the distance between objects, he determined that gravitational force is inversely related to the square of this distance, so that if the distance is increased by a factor of ten, the force will be decreased by a factor of one hundred, or if the distance is reduced by a factor of ten, the force will be increased by a factor of one hundred.

The Universal Law of Gravitation explains why we have weight, why we have an atmosphere, and how the dynamic equilibrium of gravitational forces makes our solar system work. The law provides space scientists with the mathematics to determine the trajectories needed to probe the furthest corners of our universe. It provided the crew of the Apollo 13 with the calculations needed to determine the capture and escape forces as well as the trajectories needed to use the gravitational attraction of the moon to save their lives and catapult their crippled spacecraft back to earth. Indeed, the space program has helped us all to better understand the distinction between mass and weight. Mass is the same throughout the universe. Neil Armstrong was Neil Armstrong on the launch pad at Cape Canaveral, where he was attracted to the earth's center with a force that was his weight on the surface of the earth. He was still Neil Armstrong as he appeared weightless in his Apollo 11 capsule en route to the moon, and he was still Neil Armstrong as he took those first unusually springy steps "for all mankind" on the moon in 1969 where he was pulled toward the moon's center with a force that is only one sixth of that on the earth's surface.

As one might expect, there is no concept of greater importance to space scientists than gravitation. So guess what a scientist at the Stennis Space Center in Mississippi suggested as a possible mechanism for linking sea-floor volcanism with El Niños.

["Who is buried in Grant's Tomb?"] Yep! Gravity—the most fundamental principle of the physical universe, the key topic in Chapter 1 of everyone's Physics 101, and the most elementary, important, unresolved, and humbling mystery of science! I mean, after all, how and why should it make any sense for a distribution of perfectly neutral rocks, as an example, which are scattered on the ground, or elsewhere, to be attracted to one another as well as to everything else in the universe. Of all the unanswered questions regarding our physical universe, gravitation may eventually be the one remaining question reserved exclusively for God!

As I began reading the letter suggesting that gravity could be an important consideration, I did so with a mixture of skepticism and humor. Sure, I had asked my colleagues to consider indirect methods by which El Niños and sea-floor volcanism might be linked; but there was nothing in the physical world more indirect than gravity. The scientist from the space center had brought me back to my roots, to my Physics 101. I decided to keep an open mind and use the Law of Gravitation to make some elementary calculations. My skepticism quickly evaporated.

The Numbers

As discussed earlier, a sustained weakening of atmospheric pressure amounting to about 2 or 3 parts in 1000 for a few months is associated with El Niños. Since pressure is a measure of the weight divided by the surface area over which the weight is distributed, a change in the gravitational force (i.e., the weight) of about 2 parts in 1000 (or about 0.2%) would explain a 0.2% change in pressure. The question then becomes: "What could happen below the high pressure cell over a site of sea-floor volcanism that could reduce the gravitational force?" Well, the water, rock, and molten lava could become lighter due to the expansion of those materials as a result of the intense heat associated with episodes of sea-floor volcanism. Yes, that's true, but would the numbers be in the 0.2% range?

A modest slab of water on the sea-floor heated 10° C would produce a reduction in the pressure of the air at the ocean's surface of about 0.4 parts in 1000, or 0.04%. If the water were heated 50° C rather than 10° C, the reduction would provide the 0.2% threshold associated with El Niños. The molten lava pouring out onto the sea-floor is actually at temperatures of more than 1000° C, and water temperatures of hundreds of degrees, possible only at the great pressures of the deep ocean, are measured along active sea-floor volcanoes.

In addition, consideration must be given to the gravitational effects of massive intrusions of molten lava just below the sea-floor. These intrusions may extend downward deep into the earth's mantle or core as they provide the energy for intense episodes of hydrothermal activity, underwater volcanism, seismicity, and sea-floor spreading along the world's mid-ocean ridge systems. Increases in the size and temperature of these intrusions as evidenced by intense episodes of sea-floor volcanism could contribute substantially to the reduction of atmospheric pressures through reductions in the force of gravity.

The surface of the earth's outer core, from which these intrusions may originate, is not unlike the surface of the sun which displays episodic bursts of especially intense energy on a cycle of about eleven years. Unlike the sun's surface, however, the earth's molten and dynamic core is covered by an "atmosphere" of solid rock, making it difficult to predict when or where its pent-up energy will eventually be released.

Earlier I had stated that the equatorial band of warm water is far from being the whole story and that a simple explanation for the cause of the warm water, consistent with accepted doctrines of oceanographers and atmospheric scientists, is that a big water pump (the southeast Pacific atmospheric high pressure cell) helps to push water from east to west across the equator. The fuel for this pump is cold air; and if that fuel is warmed, the pump will not work very well or it may not work at all. Now we must add that the rate of flow in the pump may also be diminished if the weight of the fuel (the cold air) is lessened by reductions in the force of gravity associated with sea-floor volcanism. When either warming of the air in the cell or reductions in the weight of the air in the cell due to sea-floor volcanism occurs, the water along the equator will slow or stop in its normal motion from east to west and will heat up, altering the normal dynamics of the earth's atmosphere with possibly devastating consequences.

Final Thoughts

Whether gravitational changes, direct or indirect heating, or some other mechanism will prove to be the primary link between sea-floor volcanism and El Niños remains an unanswered question. All of these explanations should be investigated. What triggers El Niños has been one of the earth's great mysteries. Scientists should not be intimidated by the potential embarrassment of being wrong in pursuing possibly false or seemingly bizarre leads associated with difficult, cross-disciplinary problems. Being a participant in the process of logically, thoroughly, and imaginatively searching for the truth is far more important than avoiding the process for fear of being labeled as foolish or wrong by stagnant orthodoxy. After all, Alfred Wegener, Doc Woollard, and Antonio Haoa were right! There are forces capable of splitting continents, *"Everything is related!"*, and Easter Island is "The Earth's Bellybutton".

It would be especially ironic, yet appropriate, if El Niños could be added to the list of mysteries solved by the plate tectonics revolution. But, as with Wegener's ideas, will it take decades for earth scientists to seriously consider a linkage to plate tectonics? If so, such a delay could have far more serious economic and human consequences than the original negligence of Wegener's ideas.

It would also be consistent with what often appears to be the mischievous character of Mother Nature, if the most fundamental law of our physical universe, which was derived by an unwanted mediocre student born on Christmas Day more than three hundred and fifty years ago, provided an explanation to one of the earth's most destructive and puzzling phenomena, which has mistakenly been associated with a Child of Peace born on that earliest Christmas Day.

"To myself I seem to have been only like a boy playing on the seashore, and diverting myself in now and then finding a smoother pebble or a prettier shell than ordinary, whilst the great ocean of truth lay all undiscovered before me." [Sir Isaac Newton]

The Prize

As simple as it may seem, the greatest reward that scientists will ever receive is the satisfaction and memories of their discoveries, adventures, and challenges; and the knowledge that integrity, dedication, and open-mindedness were brought to their job. We should all recognize that this is the prize that so many of our colleagues, past and present, have already earned, and is the prize most appropriate for future generations of scientists.

Epilogue

FROM A CHILD'S SUMMER DREAM
TO A LIFE-LONG ADVENTURE ACROSS THE PACIFIC
FROM EASTER ISLAND, THROUGH HAWAII
TO MIDWAY, WAKE, AND MARCUS ISLAND
ON TO POHNPEI AND JAPAN
IN PURSUIT OF ENDLESS SUMMERS

DRIVEN BY INSPIRATIONAL LEADERS
AND COLLEAGUES

MOTHER EARTH SLOWLY AND RELUCTANTLY
REVEALING HER SECRETS TO ME

SECRETS OF GREAT IMPORTANCE

GLOBAL WARMING, NUCLEAR PROLIFERATION,
MYSTERY CLOUDS

THE MOTHER'S HEARTBEAT AND HOW
SHE RENEWS HERSELF

CLUES TO EARTHQUAKE PREDICTION AND EL NIÑOS
ISAAC NEWTON AND "THE CHILD"
A JOURNEY TO SHARE WITH ALL
SCIENTISTS, ADVENTURERS, LOVERS OF MYSTERY
TO CONTINUE UNTIL THE END
SUSTENANCE FOR THE SOUL OF MAN

Bibliography

Earthquake Triggering

Berg, E., Sutton, G. H., and D. A. Walker, Dynamic interaction of seismic activity along rising and sinking edges of plate boundaries, *Tectonophysics 39*, 559-578, 1977.

Hill, D. P., Reasenberg, P. A., Michael, A., Arabaz, W. J., Beroza, G., Brumbaugh, D., Brune, J. N., Castro, R., Davis, S., dePolo, D., Ellsworth, W. L., Gomberg, J., Harmsen, S., House, L., Jackson, S. M., Johnston, M. J. S., Jones, L., Keller, R., Malone, S., Munguia, L., Nava, S., Pechmann, J. C., Sanford, A., Simpson, R. W., Smith, R. B., Stark, M., Stickney, M., Vidal, A., Walter, S., Wong, V., and J. Zollweg, Seismicity remotely triggered by the magnitude 7.3 Landers, California, earthquake, *Science 260*, 1617-1623, 1993.

Monastersky, R., Enigmatic tremors erupt across West, *Sci. News 142*, 74, 1992.

Walker, D. A., Yearly seismic energy release: world totals versus ridge system totals, *Science 193*, 886-888, 1976.

Walker, D. A., Temporal clustering of large rare earthquakes, *J. Phys. Earth 44*, 205-213, 1996.

Easter Island

Conniff, R., Easter Island unveiled, *Nat. Geographic 183-3*, 54-79, 1993.

Heyerdahl, T., *Aku-Aku*, Rand McNally, Chicago, 1958.

El Niño

Canby, T. Y., El Niño's ill wind, *Nat. Geographic 165-2*, 144-183, 1984.

Graham, N. E., and W. B. White, The El Niño cycle: a natural oscillator of the Pacific ocean-atmosphere system, *Science 240*, 1293-1301, 1988.

Kerr, R. A., El Niño metamorphosis throws forecasters, *Science 262*, 656-657, 1993.

Leybourne, B.A., A tectonic forcing function for climate modeling, *Eos 77*, supplement, W8, 1996.

Monastersky, R., Defying predictions, El Niño still simmers, *Sci. News 143*, 292-293, 1993.

Monastersky, R., El Niño gathers steam in the Pacific, *Sci. News 152*, 75, 1997.

Monastersky, R., Pacific warmth augurs weird weather, *Sci. News 151*, 316, 1997.

Ramage, C. S., El Niño, *Sci. Amer. 254-6*, 76-83, 1986.

Sea animals stranded by El Niños, *Sci. News 149*, 11, 1996.

Trenberth, K. E, Branstator, G. W., and P. A. Arkin, Origins of the 1988 North American drought, *Science 242*, 1640-1645, 1988.

Walker, D. A., More evidence indicates link between El Niños and seismicity, *Eos 76*, 33, 34, 36, 1995.

Walker, D. A., Seismicity of the East Pacific Rise: Correlations with the Southern Oscillation Index?, *Eos 69*, 857, 865-867, 1988.

Walker, D. A., Seismic predictors of El Niño revisited, *Eos 80*, 281,285,1999.

Global Warming

Broecker, W. S., Chaotic climate, *Sci. Amer.* *273-5*, 62-68, 1995.

Johnson, R. H., Synthesis of point data and path data in estimating SOFAR speed, *J. Geophys. Res. 74*, 4559, 1969.

Karl, T. R., N. Nicholls, and J. Gregory, The coming climate, *Sci. Amer. 276-5*, 78-83, 1997.

Matthews, S. W., Is our world warming?, *Nat. Geographic 178-4*, 66-99, 1990.

Monastersky, R., Health in the hot zone, *Sci. News 149*, 218-219, 1996.

Monastersky, R., The case of the global jitters, *Sci. News 149*, 140-141, 1996.

Munk, W. H., and A. M. G. Forbes, Global ocean warming: An acoustic measure? *J. Phys. Oceanogr. 19*, 1765-1778, 1989.

Raloff, J., How climate perturbations can plague us, *Sci. News 148*, 196-197, 1995.

Walker, D. A., Repeat of thirty-year-old experiment could prove or disprove global warming, *Eos 77*, 191, 1996.

Kaitoku

McKenna, D. L., and D. A Walker, Mystery cloud: additional observations, *Science 234*, 412-413, 1986.

Stone, S. C. S., An unsolved unsettling mystery, *Honolulu 20-3*, 52, 53, 75, 76, 1985.

Walker, D. A., McCreery, C. S., and F. J. Oliveria, Kaitoku Seamount and the mystery cloud of 9 April 1984, *Science 227*, 607-611, 1985.

Nuclear Explosions on Hydrophones

McCreery, C.S., and D. A. Walker, Spectral comparisons between explosion P signals from the Tuamotu Islands, Nevada, and Eastern Kazakh, *Geophys. Res. Lett. 12*, 353-356, 1985.

McCreery, C.S., Walker, D. A., and G. H. Sutton, Spectra of nuclear explosions, earthquakes, and noise from Wake Island bottom hydrophones, *Geophys. Res. Lett. 10*, 59-62, 1983.

Walker, D. A., Hydrophone recordings of underground nuclear explosions, *Geophys. Res. Lett. 7*, 465-467, 1980.

Plate Tectonics

Takeuchi, H., S. Uyeda, and H. Kanamori, *Debate About the Earth*, Freeman, Cooper & Co., San Francisco, 1967.

Davidson, K., and A. R. Williams, Under our skin: hot theories on the center of the earth, *Nat. Geographic 189-1*, 100-111, 1996.

Submarine Volcanism

Bonatti, E., The earth's mantle below the oceans, *Sci. Amer. 270-3*, 44-51, 1994.

Cartographic Division, The earth's fractured surface, *Nat. Geographic 187-4*, map, 1995.

Dziak, R. P., Fox, C. G., Embley, R. W., Lupton, J. E., Johnson, G. C., Chadwick, W. W., and R. A. Koski, Detection of and response to a probable volcanogenic T-wave event swarm on the western Blanco Transform Fault Zone, *Geo. Res. Lett. 23-8*, 873-876, 1996.

Hammond, S. R., and D. A. Walker, Ridge event detection: T-phase signals from the Juan de Fuca Spreading Center, *Mar. Geophys. Res. 13*, 331-348, 1991.

Hey, R. N., Johnson, P. D., Martinez, F., Korenaga, J., Somers, M. L., Huggett, Q. J., LeBas, T. P., Rusby, R. I., and D. F. Narr, Plate boundary reorganization at a large-offset, rapidly propagating rift, *Nature 378*, 167-170, 1995.

Kunzig, R., Can Earth's internal heat drive ocean circulation?, *Science 252*, 1620-1621, 1991.

Lutz, R. A., and R. M. Haymon, Rebirth of a deep-sea vent, *Nat. Geographic 186-5*, 114-126, 1994.

McCreery, C. S., Oliveira, F. J., Walker, D. A., Hamada, N., and J. Talandier, Submarine volcano, *Eos 70*, 1466, 1989.

McCreery, C.S., Walker, D. A., and J. Talandier, Hydroacoustics detect submarine volcanism, *Eos 74*, 85-86, 1993.

Walker, D.A., and S.R. Hammond, Spatial and temporal distributions of T-phase source locations on the Juan de Fuca and Gorda Ridges, *Eos 71*, 1601, 1990.

Others

Hapgood, F., The quest for oil, *Nat. Geographic 176-2*, 226-259, 1989.

Kroenke, L. W., and D. A. Walker, Evidence for the formation of a new trench in the Western Pacific, *Eos 67*, 145-146, 1986.

Patterson, C. B., At the birth of nations: in the far Pacific, *Nat. Geographic 170-4*, 460-499, 1986.

The World Book Encyclopedia, World Book Inc., Chicago, 1995.

Walker, D. A., Deep ocean seismology, *Eos 65*, 2-3, 1984.

Walker, D. A., and C. S. McCreery, Deep ocean seismology: Seismicity of the Northwestern Pacific Basin, *Eos 69*, 737, 742-743, 1988.

9 780738 837222

90000>